Electricity for the 4-H Scientist

Idaho Agricultural Extension Service

Bulletin 396, June, 1962

Eric B. Wilson

Classical Prints

This edition published in 2022

ISBN : 9789356535534

Contents

HOW TO USE THIS BOOK IN FULFILLING THE GOALS OF THE 4-H ELECTRIC PROJECT FOR THE FIRST AND SUCCEEDING YEARS

The minimum goals for credit in the 4-H Electric project vary according to the 4-H member's age and the number of years he or she has taken the electric project. For example, if you are a 4-H member beginning the 4-H Electric project at the age of 10, you will not be required to earn as many credit points as a 14-year-old 4-H member beginning the 4-H Electric project. However, if you are a 12-year-old in your second year of electricity you must earn as many credit points in that year as a 14-year-old does in his or her first year.

Each lesson or goal has been designated a certain number of credit points. These are shown near the title of each lesson or goal. You decide on the lessons you want to study, list them, and add up the credit points.

For a full year's 4-H project credit, the total of your credit points should be at least as many as shown in the following table:

Examples of reading the table below are as follows: (a) An 11-year-old member is required to complete 13 credit points the first year, (b) A 14-year-old is required to complete 17 credit points his first year, (c) A 14-year-old taking the electric project for the third year must complete 16 credit points that year.

We recommend that, if you are taking the 4-H Electric project, you start with the first lesson in the book and go on through to the back of the book in advanced years. But you may skip the less important or less interesting parts so long as you learn the basic lessons. A way to find out whether you know the basic lessons is to read them through and try to answer all questions under the heading "What Did You Learn." If you can answer these questions you may not wish to spend the time doing the things listed under "What To Do."

Minimum Number of Credit Points Required for Each Year's Work in the 4-H Electric Project

4-H Member's Age	4-H Member's Year in 4-H Electric Project			
	1st Year	2nd Year	3rd Year	4th or Later Years
10-11	13	15		
12-13	15	17	19	20
14-15	17	19	21	21
16 & over	19	21	21	21

This system of credit points makes it possible for you to do the things you want to do with electricity and get credit for them in the 4-H Electric project.

4-H Electric, Division I

LESSON NO. B-1

Credit Points 3

GETTING ACQUAINTED WITH ELECTRICITY

Electricity serves you best when you understand how it works and use it properly. As a 4-H member, you should know about electricity and help to show others the way to obtain its tremendous work-saving benefits as well as how to use it with safety.

A good way to think of electricity is to compare it with water. It acts a lot like water. However it is made of tiny parts of atoms called electrons. When there are more than the normal number of electrons in anything, it is said to be negatively charged; when there is a shortage of electrons, it is positively charged. As water flows downhill, "seeking it's level," electrons flow from negative to positive, seeking to "balance" the charge.

Electrical Conductors

Even if you're never going to repair a lamp or make a chick brooder, you should know about conductors and insulators. This is because you happen to be a fairly good conductor of electricity. Electricity will pass easily through you to other conductors—the ground, for instance. When this happens you may get a shock, burn, or serious injury. But it doesn't ever have to happen, if you learn to understand your friend, electricity.

Silver, copper, iron, aluminum and many other metals are very good conductors. Water, acids, and salts are too. Electricity passes over or through them very easily. Like water pipes, the larger the conductor, the more electricity it can carry. When conductors are too small for the amount of electrons trying to move over them, they get hot, melt, may start fires. That's why wire size is important.

Electrical Insulators

Insulators are the opposite of conductors. Electricity has trouble passing through some materials. Rubber, most plastics, dry wood, oils and glass are some of the good insulators. It's the amount and kind of insulation that counts. If it has enough force, electricity can pass through just about anything—even jump gaps!

Electricity, like water, flows along the easiest paths. It is always trying to get to the ground. The earth attracts it. It stays on the wires unless a person, a wet branch, or some other conductor gives it a path to the ground. Do not touch any wire which might be carrying electricity.

Play It Safe

If you should touch a "hot" wire accidentally and are standing on a dry piece of wood, the conducting pathway to the ground is not good and the electricity may keep running along its wire. But do not touch some other conductor with another part of your body. This would complete a circuit through your body and would be very dangerous. Always make sure there is plenty of good insulation material or plenty of distance between you and anything which might be carrying electricity.

Remember, too, insulation is of little use when it is wet. Dew, mist, rain, condensation, a damp floor can change the whole picture. If you understand electricity and how it acts, you'll be safe enough, because you won't take chances or expose yourself to injury.

Electrical Terms

Alternating Current—Usually referred to as "AC," alternating current is current which reverses its direction of flow at regular intervals, 60 times a second.

Direct Current—"DC" current flows only in one direction. Battery current is DC.

Ampere—Amperes are units by which the rate of flow of electrical current (electrons) is measured. An ampere is 6.3 billion electrons passing one point in a circuit, in one second. This compares with the way the flow of water is measured in gallons per second.

Volts—A volt is a unit to measure the tendency of electrons to move when they are shoved. Voltage is the amount of "push" behind the electrons. It's like water pressure in a pipe. Home power lines carry 115 volts (110 to 120 volts). For appliances such as electric stoves, washers and driers, a second 115-volt line should be added, giving 230 volts (220 to 240 volts).

Watts—Watts equal volts times amperes. Light bulbs, electric irons and other appliances are usually marked with the voltage they require and the number of watts.

Kilowatts—Your electric bill usually reads in kilowatt hours. A kilowatt is 1000 watts. A kilowatt hour equals 1000 watts used for 1 hour. One kilowatt equals about 1-1/3 horsepower. A kilowatt is usually indicated by "kw" and a kilowatt hour by "kwh."

Circuits—A closed circuit is one in which the electricity is flowing, lighting a light, running a motor, or some other appliance. The circuit runs all the way from the place the electricity is being generated to your home, through the appliance or light bulb, and back to the generator.

Circuits are opened and closed by switches. When the circuit is opened, the electricity stops at the switch. Before working on a switch, socket, fuse, or any part of the wiring be sure to open the main switch. The main switch is usually at the fuse box or near it. Appliances should be disconnected when you work on them. Everyone in the family should know where the main switch is so it can be pulled in case of accidents, fire, flood, or windstorm damage.

Fuses and Circuit Breakers—These are the safety valves of your electrical system. The different electrical circuits in your home are meant to carry only certain amounts of electricity. Some carry only 15 amps, others can carry 20 or more. They are marked to show capacity.

When a fuse burns out or a circuit breaker opens, look for an overload of lights and appliances on the circuit before you try to replace the fuse or close the circuit breaker. Without these safeguards, the overloaded electric line will heat up and may start a fire. Even if no fire starts, electricity will be wasted and the homeowner will be paying for electricity that's doing no good.

Remember: If you ever have to replace a fuse, pull the main switch first. Keep a flashlight handy in your house. It seems that fuses usually blow at night, and it doesn't pay to stumble or fumble around electric wires in the dark.

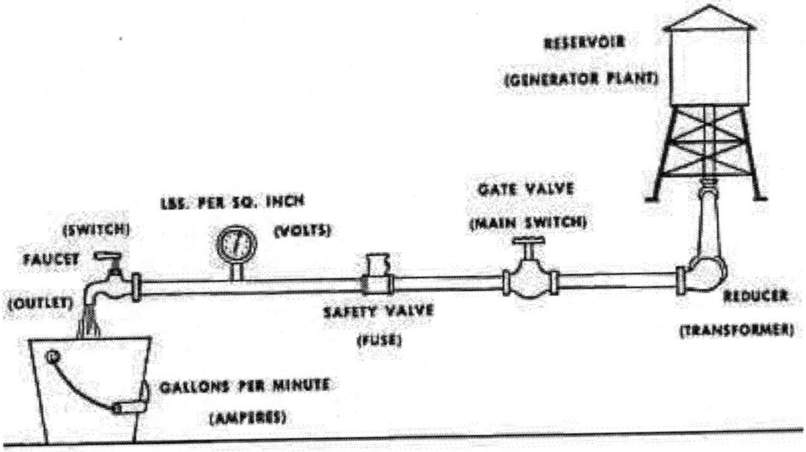

WHAT TO DO: Make A Circuit Board

So that you can show others how electricity travels from here to there, and how it behaves under different conditions, make an electric circuit board.

Materials Needed:

Piece of 3/4" board about 4" x 6"
l-l/2-volt No. 6 dry cell battery
Two pieces of bell wire, each 24" long, one black, one white
Two 10-penny box nails (3")
Three 3-penny box nails (1")
Two small screws or carpet tacks
Two 2-inch rubber bands
Two miniature sockets with solder terminals
Two l-l/2-volt flashlight bulbs

Tools Needed:

Ruler, pencils, hammer, pliers or vise.

Making the Board:

1. Lay out the board with a pencil and ruler as indicated in Figure 1.
2. Bend the three-inch nail as shown in Figure 2, using pliers, vise and hammer.
3. Pound the one-inch nails into the board for a half-inch at points A, C, and D. Use the three-inch nail to make a hole a half-inch deep at B. Put the

crank nail in this hole and pound in a little farther. Attach the lamp socket brackets at E and F. Stretch the rubber band as in Figure 3.

4. Lay out the electricity path, the circuit (Figure 3). Use the black wire for the positive side of the circuit (the center pole of battery). Twist it around the switch crank B, and the center pole of battery. Run another piece to the outside terminal of bulb socket at E. Run white piece to negative pole of battery from the other terminal at E.

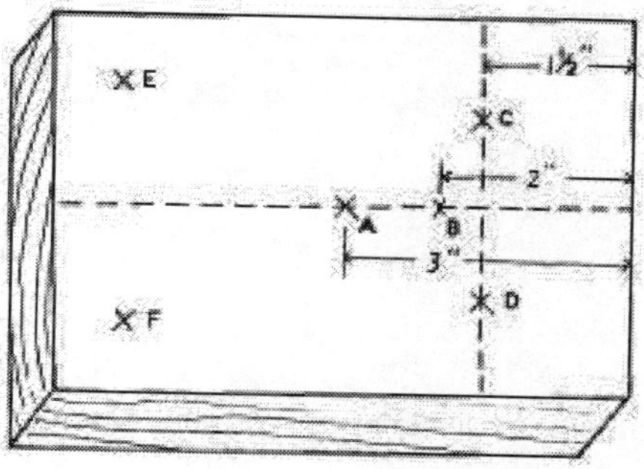

Figure 1 (Circuit Board)

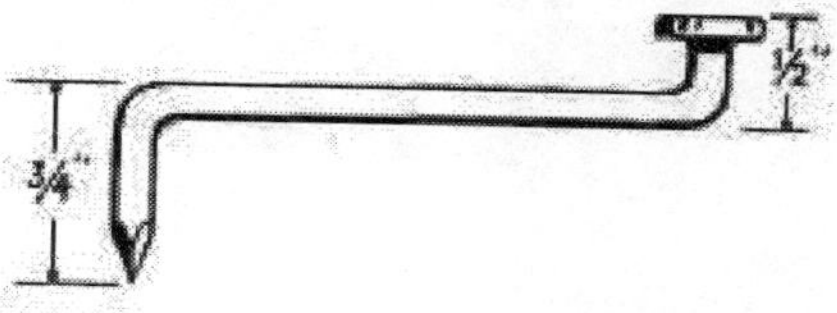

Figure 2 (Switch)

5. Close the switch. The rubber band should hold the switch nail tightly against nail at C. Does the bulb light? __________ If it doesn't, check the connections.

Now you have a circuit—a closed circuit when the electricity runs all the way from the positive pole to the negative pole. The black wire is the hot side, the live wire, because it carries the full load of the battery up to the bulb.

Remember, battery current is direct current, DC. In the case of alternating current, AC, such as most homes and buildings use, the electricity flows in first one direction and then the other.

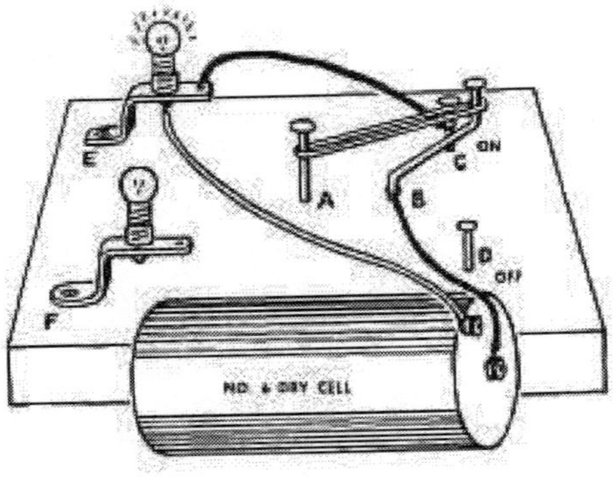

Figure 3 (Closed Circuit)

Parallel Wiring

To make this circuit hookup, attach another white wire to the negative pole of battery and a terminal of the second flashlight bulb. Run a black wire from the other terminal to the switch terminal at C (Figure 4). Close switch. Both bulbs will light.

Trace the circuit. Electricity is going equally to each bulb, the same amount that went to the single bulb. The difference is that the battery will last only half as long. It's like a pail of water with two open spigots. The pail empties twice as fast as it would with just one spigot open. This type of wiring is called parallel wiring. If one bulb is unscrewed, the other will stay lit.

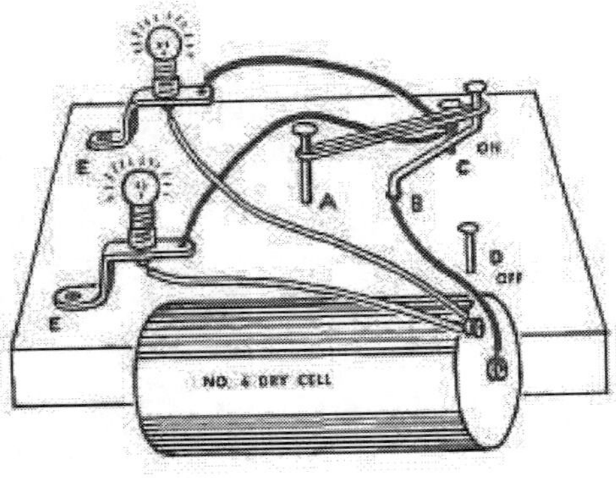

Figure 4 (Parallel Wiring)

Series Wiring

To do this, run the negative wire to one terminal of the second bulb and attach a wire from the other terminal to a terminal of the first bulb. The other terminal connects with the switch at C (Figure 5). This is series wiring. If one bulb is unscrewed, the other will fail to light because the circuit is broken for both. Anything that breaks the circuit has the effect of opening the switch.

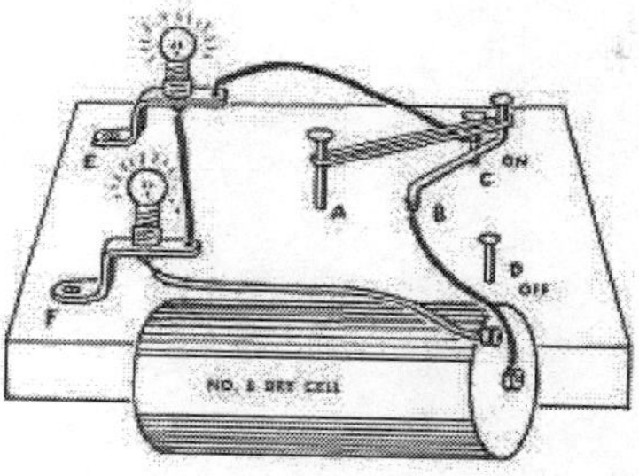

Figure 5 (Series Wiring)

Show there is a circuit through the bulb by screwing and unscrewing it. Also, "jump" the socket by running the wire from C to the other terminal of the bulb at E while it is unscrewed. Bulb at F will light. Trace this circuit.

SUGGESTED DEMONSTRATIONS

Using the Circuit Board, you can give many demonstrations of the way electricity flows, works and behaves.

Water And Electricity

To help others understand electricity better, draw a water system on an electric circuit board paralleling the circuit. For the battery show a water tank, pipes instead of wires, faucets instead of switches. Somewhere on the board paste a comparison of electrical terms with terms used in describing water, such as the following:

Wire equals Pipe
Volts equal Pressure
Amperes equal Rate of Flow - gallons per second
Watts equal Pressure times Rate of Flow
Switch equals Faucet
Current equals Flowing Water

Show how to figure the wattage that a circuit protected by a 15 ampere fuse can handle. Do it with actual things or cut-out pictures of light bulbs, irons, toasters, coffee-makers, etc.

You know that Amperes times Volts equal Watts. If the voltage is 115, a 15 amp circuit can handle 115 volts times 15 amps, or 1725 watts.

The name plates on electric motors indicate the amperage at full load. You can convert this to watts, of course, by multiplying amperage by the line voltage. Motors require an additional amount of electricity when they start. You need to allow for this fact, so fuses will not blow or circuits trip when a motor is turned on. You will learn more about this when you study electric motors.

For More Information

Your leader has many other sources of information about electricity and demonstrations you can perform. Ask him. Also, libraries have many books about electricity and its history, which are very interesting and useful. Maybe you can find an electrician, someone from your power supplier, or an equipment dealer who will talk to your club on electricity or electrical safety.

What Did You Learn?

(Underline the correct answers then discuss in the group.)

1. In a water pipe system water flows. In an electrical circuit (electrons) (atoms) (charges) flow.

2. Electricity or electrons flow (easier) (harder) (about the same) in a conductor than in an insulator.

3. Rubber is a good (conductor) (insulator) (ground).

4. The most common material used as an electrical conductor is (glass) (silver) (copper).

5. The unit of electrical pressure or push is the (ampere) (volt) (watt).

6. The rate of flow of electricity is measured in (gallons) (amperes per minute) (amperes).

7. Volts times amperes equals (watts) (kilowatt hours) (alternating current).

8. A dry cell battery (stores) (makes) (uses) electrical energy.

9. In a parallel circuit the electricity has (one) (two or more) (no) paths to travel.

10. In a series circuit with two bulbs and a switch the bulbs are (brighter) (dimmer) (the same) as when they were in the parallel circuit.

LESSON NO. B-2

Credit Points 4

TOOLS FOR ELECTRICIANS

Who goeth a borrowing
Goeth a sorrowing
Few lend (but fools)
Their working tools

Tusser 1524-1580

Whenever a job comes up, it saves time and trouble when you have the right tools and they are all where you can find them. Electrical work takes some special tools and some everyday tools.

If you have ever watched a good electrician at work, you've seen how neatly he stores his tools in a box so every one of them is handy. When a lineman climbs a pole, he has his regular tools in a holster on his belt. Special tools are kept in a box in racks in the repair truck, all ready for instant use. Wouldn't you like to have electrician's tools all handy, ready for use, and know how to use them properly?

Basic Tools for Electrical Work

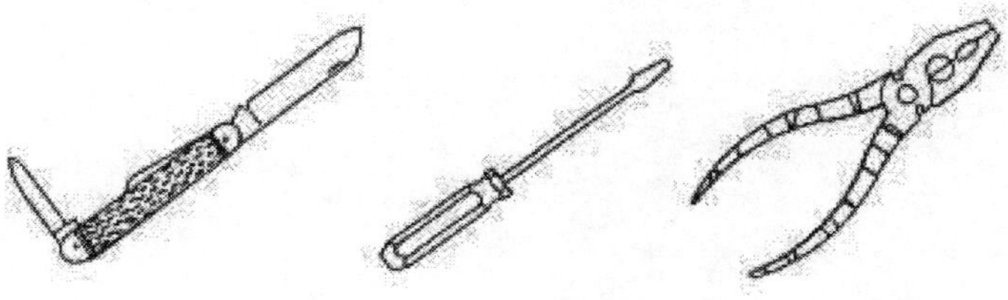

Knife

A good knife with a sharp blade is one of the most useful tools. A camper's or electrician's type knife is probably best because it has other useful parts besides the cutting blades—a screwdriver or punch, for instance. Of course, you'll never use the cutting blades as a screwdriver. This knife should be kept clean, dry, sharp, and free from rust. Put a little oil on the joints from time to time. Remember, "Never whittle toward you and you'll never cut yourself."

Pliers

A pair of electrician's pliers should be part of your kit. Wrap the handles with plastic insulating tape. Even though you're not going to work on "hot" electric lines, it pays to play safe. Later on, as you learn more about electricity, you'll want a pair of needle-nose pliers for the fine work.

Screwdrivers

You'll want a screwdriver which has true corners. A 4 to 6 inch plastic handled screwdriver with a narrow blade is best. You'll probably need more than one size to fit the various size screws you'll be turning.

Screwdrivers are easily damaged if you try to use them as chisels and pry bars, or use them in screw slots which are too large for the blade.

You can be hurt by the screwdriver if you try to screw or unscrew things you are holding in your hand. Keep your free hand away from the end of the screwdriver. Place the work on a bench or where it can be handled easily.

Soldering Iron

A good 100 to 250-watt electric soldering iron will be useful. Later on you may want to buy a soldering gun, but unless you are doing a lot of soldering it won't be necessary. A supply of resin-core electrician's solder will be needed. Acid-core solder reacts with copper and in time causes a bad splice.

Tape

Once it was necessary to use two types of tape on splices—rubber tape with friction tape over it. Now there is a plastic tape on the market which takes the place of both and has good insulating quality. It is called electrical tape, or plastic tape, and resists water, oils (which would damage rubber tape), and acids. You'll need a lot of tape in your electrical work, so keep a roll on hand.

Other Tools and Equipment

As you go along in electrical work, you'll be adding tools and other equipment, such as a trouble light and maybe an ammeter or voltmeter. Other tools you'll want to add will be a Phillips screwdriver, open end wrenches, a crescent wrench, small hack saw, hand drill and bits.

You'll also be using some regular carpenter's tools such as hammers, saws, and so on. Unless you use them frequently, you don't need to keep them in your electrical kit.

It's a good idea to start acquiring a supply of electrical parts—lengths of wire, fuses, switches, sockets, plugs, and other items that will come in handy. There are parts you can salvage from old lamps, motors, and other equipment. Such a collection can be a real treasure chest when you need a part in a hurry. But be sure to throw away all faulty parts.

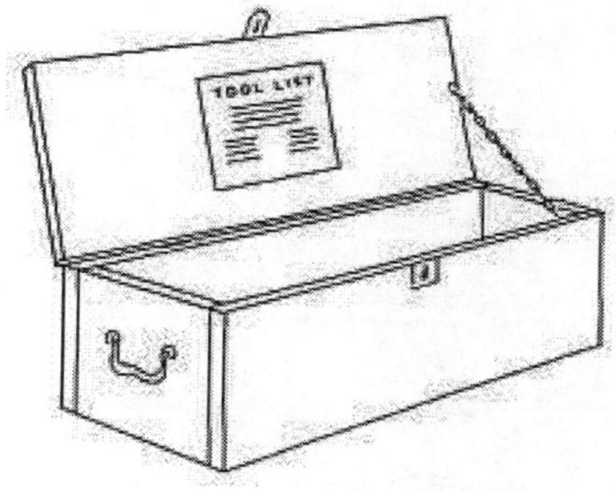

Figure 1. Completed tool chest.

WHAT TO DO: Build a Tool Chest

To keep your tools always ready for use, a tool chest will be very handy. It's the 4-H way to work. You'll be surprised how much easier it makes a job when you have your tools, various parts and repair equipment all in one place. You can make the chest (Figure 1) with a saw, plane, screwdriver, pencil, ruler or carpenter's square, and hammer.

Materials You'll Need:

A piece of lumber 1" by 10" by 8 feet long. (1" lumber is actually only 3/4" thick—this is the thickness you'll be working with.)

2 small hinges, with wood screws
1 small hasp, with wood screws
2 small handles with wood screws, or one large handle
1 small chain, 10" to 12" long
Some No. 6 penny finishing nails or wood screws about the same length

Making The Chest:

1. Cut your lumber into the following pieces:

1 piece 10" x 18" for top

1 piece 8-1/2" x 16-1/2" for bottom

2 pieces 6" x 8-1/2" for two ends

2 pieces 6" x 18" for front and back

2. Lay out pieces as shown in Figure 2.

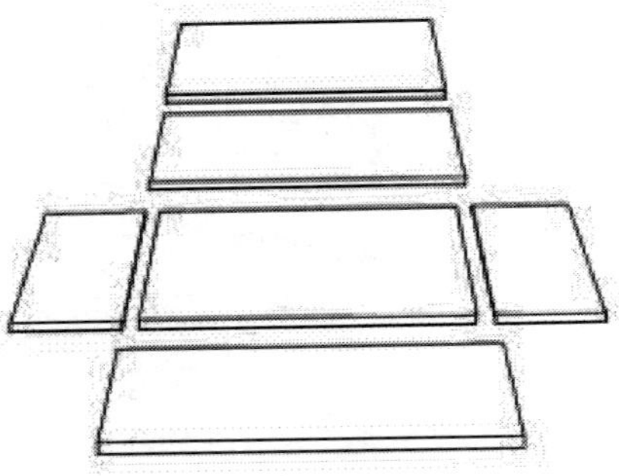

Figure 2.

Then, set up the two end pieces and nail to bottom section. Refer back to Figure 1 as you go along to see that box is shaping up as shown. Nail the front and back sections to the ends along the bottom. Wood screws can be used instead of nails.

3. Lay the top in place and attach hinges to the back side, about two inches in from each end.

4. Attach one part of hasp to the top, and the other part to front board in center. Fasten the handles to each end.

5. Attach chain to the top and front so the top will stay open when chain is fully extended.

Now you can invent your own improvements for your chest. You can paint it, put your name on it, and your club emblem and name if you wish. You can put a rack on the inside of the cover to hold your work sheets and other booklets and materials. You can install special slots or straps to hold each tool in its place along the sides of the box. Maybe you will want to put some partitions in the box to separate various electrical equipment such as wires, fuses, switches, and plugs.

A Working Kit

An accessory which you may want to add to your tool chest is an apron or holster to wear when you are moving around on the job. An apron can be made of a size of cloth about 18 by 20 inches. It should be folded up from the bottom, and sewn to fit the number and size of tools you have. Figure 3 shows such an apron.

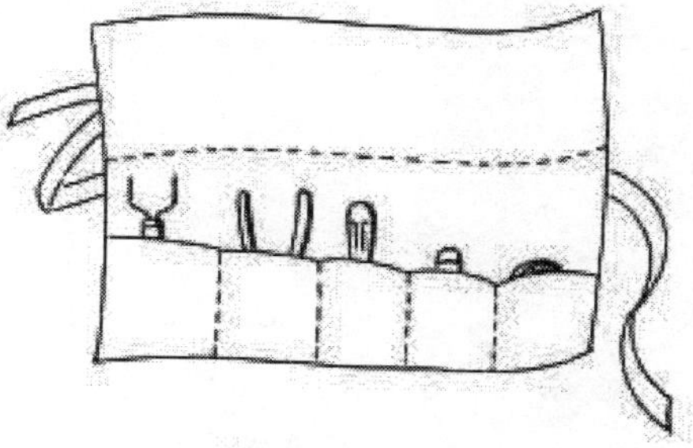

Figure 3. Apron

You can make a lineman's holster in the same way, using plastic or soft leather. Merely make belt loops by cutting on the dotted lines. A snap fastener will hold the flap over the tools so they won't fall out.

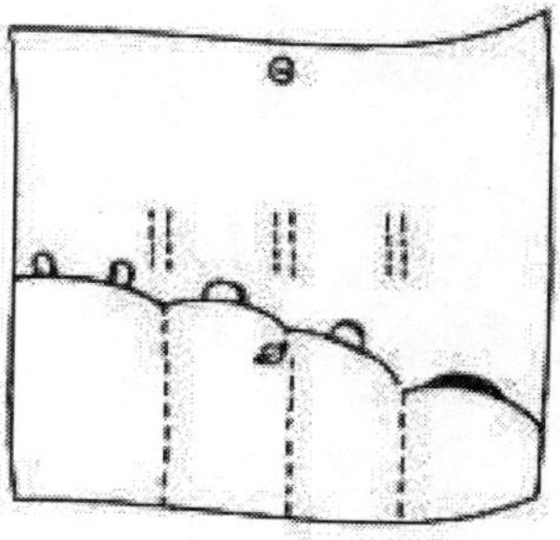

Figure 4. Lineman's Holster.

Demonstrations You Can Give

Show and tell others the proper handling, care and use of tools.

Show and tell how to build an electrician's tool kit.

For Further Information

Ask your power supplier or an electrician to tell the club about the various tools of the electrician's trade and demonstrate them. Ask your leader how to get exhibit material or information about electrical tools and their use and then tell the club about them.

LESSON NO. B-3

Credit Points 4

REWIRE A LAMP—BE A LAMP DETECTIVE

The Line-Up Of Lamp Suspects

One of the duties of a law officer is to prevent crime. It's that way with the lamp detective. You can become one. In the average home there are lamps about to commit the crime of shocking people, starting fires, and stealing electricity. Some are refusing to do their job well and some are no-goods, sitting in closets or attics, doing nothing. You can put these lamps to working again safely and well. Become the lamp expert in your family.

What's In A Lamp?

A lamp gives light for comfortable and convenient use in the home. It consists normally of a stand, switch, cord, lampshade holder, and shade. Some lamps have diffusing bowls which reduce glare and shadows.

The most common fault found in an old lamp is in the cord, but sometimes the switch or the wiring in the lamp is bad. Look over all the lamps in your home and find the ones needing to be fixed.

WHAT TO DO—Rewire A Lamp

Somewhere around your house you can probably find a lamp that is no longer used or needs repairing. You can make it useful again and at the same time learn how to wire a lamp.

Materials Needed:

Tools: Pocket knife, small or medium screwdriver, and pliers (electrician type is best).

New Lamp Cord: For each lamp to be rewired, you'll need 6 feet of cord plus the length of wire within the lamp stand. Lamp cord wire comes in two sizes, No. 18 and No. 16 AWG (American Wire Gauge). No. 18 is smaller than No. 16, but is adequate for most lamps. Cords are made with surface coverings of several different materials: braided cotton, rayon or silk, and molded rubber or plastic. Braided cord is decorative, but rubber or plastic is easier to work with and is usually more desirable.

Switch: If the switch is bad, get a new one. Socket switches are made with push-through, turn-knob, or pull-chain controls. The pull-chain type is seldom used on modern table or floor lamps. Your lamp may have a separate push-switch in the base. In this case, get the same kind for replacement. Some switches are "3-circuit" switches for use with high, medium, and low-light bulbs.

Plug: Plugs are made of various materials, mostly hard rubber or molded plastic. Some have a shank or handle for better grasping. This type is more desirable. The plug on the old cord may be good, and if so, may be used on the new cord.

How To Do It:

1. If the plug on the old cord is good and you plan to use it, remove it from the old cord.

2. Measure and cut a new lamp cord equal to the length of the cord within the lamp, plus 6 feet.

3. Pass one end of the new cord through the center of the plug. Strip 2 inches of the fabric insulation off cord, or in case of a rubber cord, split cord back two inches. Be sure no bare wire shows in long split section (Figure 1).

4. Use knife to strip insulation off wire for 3/4" on end of each cord. Be careful. Don't cut yourself. Don't cut wires. Use a light touch, slope the knifeblade and slice with knife edge away from you (Figure 1).

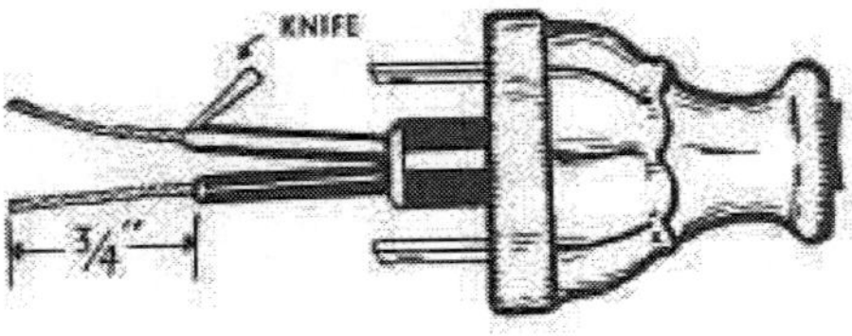

Figure 1 (Ready to Wire Plug)

5. Twist exposed strands of each wire tightly to make a good conductor, and place each conductor around its proper terminal in the direction in which the screw tightens (Figure 2).

6. Tighten screws on terminal posts. Pull cord until slack is out. Lay aside until ready to attach to lamp.

Figure 2 (Attaching Cord to Plug)

7. Remove lamp shade, shade-holder, bulb, and diffusing bowl, if there is one.

8. Separate the metal shell of socket from its cap by pressing on shell at place marked "press," and pull socket from cap.

9. Pull on socket body to get some slack in lamp cord. Loosen screws and detach cord. Pull cord out through base of lamp. You can splice new cord to the old one and use the latter to "string" the new wire.

10. Pass the new cord up through the lamp base and socket cap, tie a simple half-hitch knot in the cord to prevent strain on the terminals, and attach wires to the terminals on the socket (Figure 3). If there is likely to be any strain on cord, use an Underwriters' knot. Twist strands and attach wire in direction in which screw tightens.

11. Pull slack out of cord in lamp so that socket rests in socket cap, replace shell and reconnect cap. Be sure the fiber insulator is in the shell. You'll feel or hear a click when the notches in shell are locked to the projections in the cap.

12. Replace bulb, inspect carefully, and test. (In floor lamps where the cord runs through the center post and out under the base, the cord will last longer if it is fastened with tape so it doesn't rub edge of lamp base when lamp is moved.)

13. If the lamp has a porcelain socket, simply disconnect the wires at the terminals, remove the old wire and connect the new one.

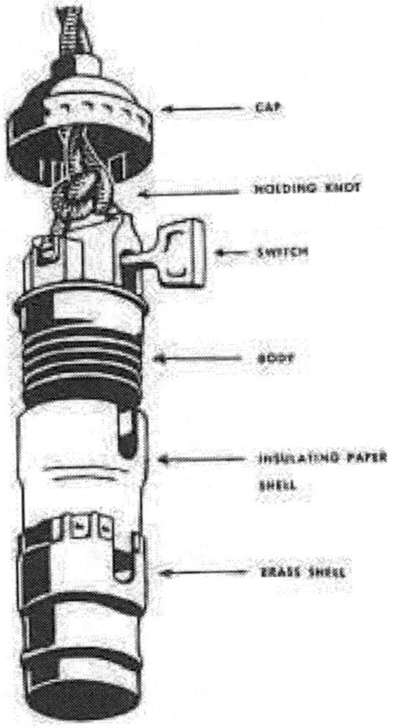

Figure 3 (Socket and Switch Assembly)

What Did You Learn?

Underline correct answers then discuss in the group. (There may be more than one correct answer.)

1. The part of the lamp that usually wears out first is (the socket) (the cord) (the plug).

2. Lamps that waste electricity are those which have (bad wiring) (frayed cords) (dirty shades or bulb).

3. To unplug a lamp you should grasp (cord) (plug) firmly and pull.

4. Wire in lamp cord usually comes in sizes 16 or 18. Size 16 is the smaller (true) (false).

5. In fastening wire around a terminal post it should go around in a (clockwise) (counter-clockwise) direction.

6. When the switch on a lamp is turned off, the electricity only goes as far as (the wall plug) (the switch).

7. An Underwriters' knot should be used (only when there is room for it in the plug) (whenever there is likely to be strain on the cord, even if you have to replace the plug with a larger one).

SUGGESTED DEMONSTRATIONS

Show how to inspect a lamp and its cord. You might tie tags on the cord and lamp at points of danger or failure—at the plug, wear points next to lamp base, bad sockets.

Demonstrate the process of repairing a lamp cord, socket and plug.

Make a board display of the parts of the lamp socket showing cord attached.

Make a display of the types of lamp cords and plugs in common use.

Using two lamps, one with clean bulb and shade, the other dusty, show how the former gives more light.

For More Information

Lamps have an interesting history. Look it up in your local library. Ask someone from your power supplier or electric dealer to talk to the club about the different kinds of lamps. Your leader has or can get additional information on lamps, if you wish.

What Did You Exhibit

What Did You Demonstrate

LESSON NO. B-4

Credit Points 3

MAKE A TROUBLE LIGHT.

A handy piece of equipment in the home and on the farm is a heavy-duty extension cord with a shielded light and a side outlet on it. When you want to work on the car or tractor in the yard at night, the trouble light is better than a flashlight. You can use it both for light and as an extension cord. It is safer than matches or a lantern, especially around the garage or barn.

It is easy to make a trouble light, and it gives you good practice in electrical work. Of course you can buy one, but you wouldn't have the fun of making it nor would it suit your needs. Trouble lights are not for permanent use—they're for emergency use and to provide light or electricity in places where they are seldom needed. When you find a trouble light being used as permanent wiring, that's the place to install an outlet.

What Size Cord?

Choose the right kind of cord. What length will be best for your various uses? A cord too long may be bothersome to use and store. What will be the heaviest load you are likely to put on the cord, in amperes? Check appliances you may want to connect to it. No. 16 wire can carry 10 amperes safely for a distance of 50 feet, while No. 18 can carry only up to 7 amperes for a distance of 40 feet. You'll want a "hard service" cord, called S, ST, or SO-type cord by electricians. Junior hard service cords, known as SJ, SJT, or SJO, are fine for lighter duty.

Cord, Plug and Guard

A rubber-handled socket should be used for safety and to withstand hard knocks. It should have a switch on it, preferably a push switch in a recess in the handle.

The connector or attachment plug should be of rubber or solid plastic and have a metal cord grip fastened to it. This grip will hold the cord firmly and prevent strain on the terminal connections.

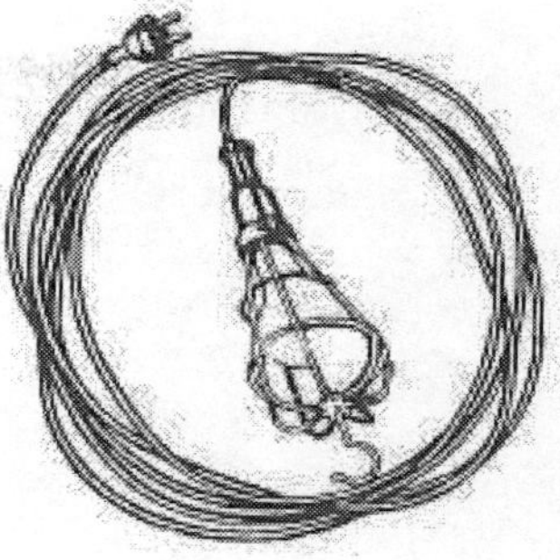

Finished Trouble Light

Get a good lamp guard. If the wire is too light, it may bend and break the bulb when hit or dropped. For the lamp itself, get a rough service lamp. An ordinary lamp won't last long with rough usage.

How to Make the Trouble Light

Tools Needed:

Your 4-H electrician's kit or screwdriver, knife and soldering iron

Materials Needed:

1. About 20 feet of 2-wire, No. 16 heavy duty (hard service)
2. A rubber-handled socket with switch and a side outlet
3. A shielded lamp guard

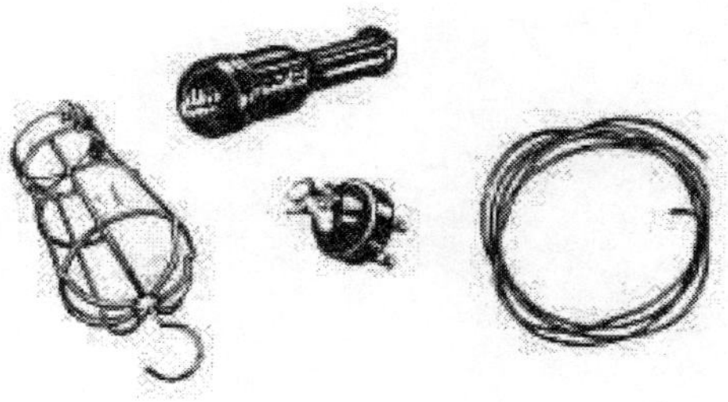

Materials Needed

4. A good connector plug cap, preferably with a clamp-type grip for the cord
5. A rough service lamp bulb
6. Solder and flux

Steps to Take:

1. Remove about 2 inches of the outer covering of cord at one end.

2. Separate the wires and cut away the filler material.

3. Remove 3/4 inch of the conductor insulation from the end of each wire and tightly twist the strands together to form a firm conductor. Be careful not to cut any of the fine wires. Ends may be soldered.

4. Slide the plug in position on the cord.

5. If there is no cord grip, tie the underwriters' knot (Figure 1). If there isn't room enough, make an "S" loop by passing the wires around the prongs before fastening them to the terminal screws as explained in the next step.

6. Loop the bare part of the wire around the screw in the direction the screw is turned to tighten (clockwise direction). This will prevent the wires from being forced out from under the head of the screw as it is tightened. Now repeat with the second wire, wrapping it around the other prong of the plug.

Connecting the socket.

1. Separate the parts of rubber-handled socket (Figure 2).

2. Prepare the other end of the cord as in steps 1, 2, and 3 above.

3. Insert the cord through the rubber handle and socket guard.

4. Tie the holding knot (underwriters' knot) as explained in Step 5.

5. Connect wires to terminal screws and assemble the rubber-handled socket.

6. Screw in the rough service lamp and test your cord.

7. Put the shielded lamp guard on the socket and tighten the holding clamp until it is firmly in place. You are now ready to use or demonstrate your trouble light.

8. After you've made your trouble light, decide on a good place to keep it where it will be handy for use. Loop it carefully and hang it over a wooden dowel rather than a nail. It will last longer.

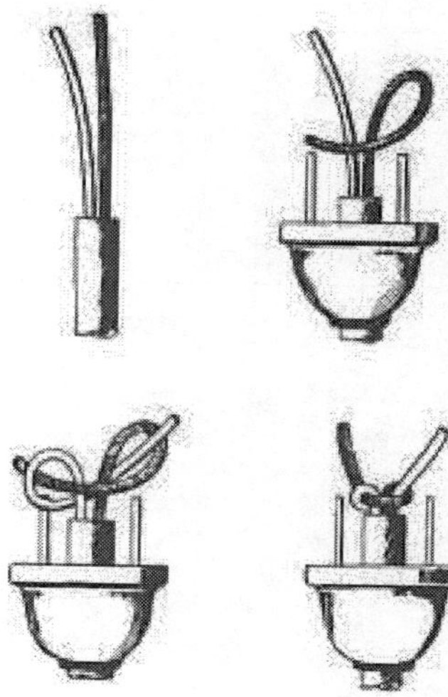

Figure 1 Tying an Underwriter's Knot

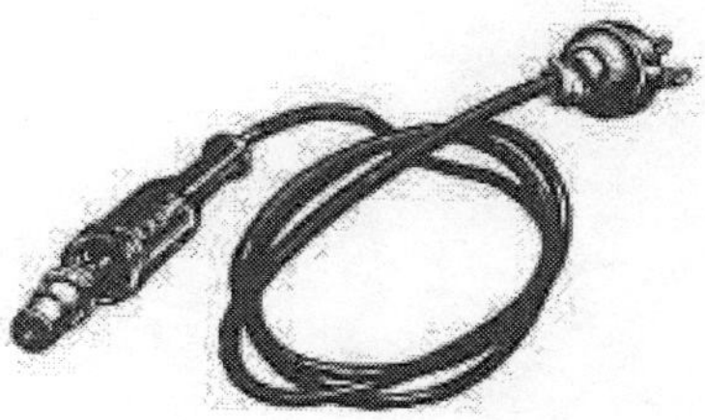

Figure 2 Disassembled Light

What Did You Learn?

(Underline correct answer)

1. A Junior Hard Service Cord is known as an (SO-Type) (SJO-Type) cord.

2. You disconnect a cord by (jerking it from the socket) (grasping plug and pulling it out).

3. Brass sockets are unsafe because (they break too easily) (the exposed metal can cause short circuits).

4. Rubber-covered cord is safer for emergency cords than fabric because (it will stretch) (it will insulate and protect the wires inside).

5. In a trouble light (any kind of bulb will do) (a rough service bulb is best).

Ideas for Demonstrations and Exhibits

1. Show how to make your trouble light and a method of storing it.

2. Show a safe trouble light, and an unsafe trouble light with danger points marked.

3. Show cutaway pieces of different types of cord.

For More Information

Ask your power supplier, county highway engineer, police official or leader to tell you about various types of portable emergency lights and their uses.

LESSON NO. B-5

Credit Points 5

WHAT MAKES MOTORS RUN

What makes an electric motor run? Can you make an electric motor that will run? Certainly you can, and by doing so you'll learn why it runs. It won't be mysterious any more and you'll be ahead of all the millions of people who use motors every day and never know why or how the motor converts electrical energy into useful power.

Motors Are Magnets

You know how one end of a compass needle always points to North. No matter how you turn the compass, the same end of the needle always swings to the North. The earth itself and that small compass are both magnets (Figure 1). Each has a North pole and a South pole. Around the poles of each there are magnetic fields, invisible lines of force that attract and repel.

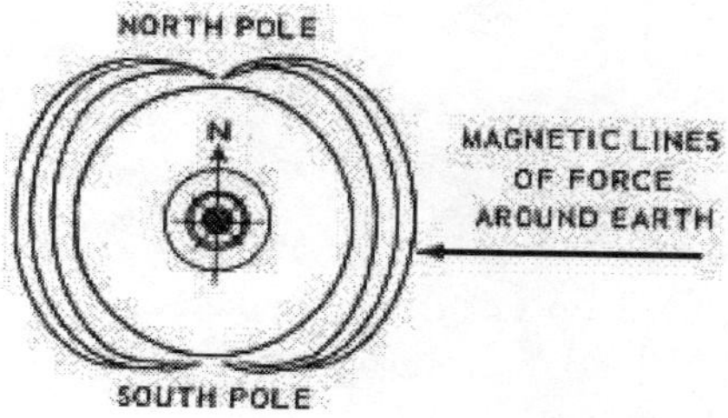

Figure 1. The same end of the compass needle always points to the earth's magnetic North Pole.

The N poles *repel* each other and so do the S poles. The N and S poles *attract* each other. In other words, opposite poles attract; poles that are alike repel each other.

Lay 2 bar magnets on a table side-by-side. If both N poles are at one end, they'll repel each other and almost flip around until there's a N pole lying next to a S pole (Figure 2).

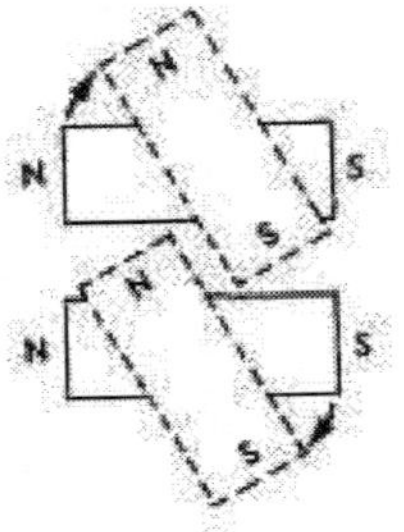

Figure 2. Small bar magnets laid side by side move so that the North pole of one is near the South pole of the other.

Now suppose we place one of the bar magnets on the table. The other, we'll fix on a pivot so it can spin around. This one we'll move so its N pole almost touches the fixed magnet's N pole. As soon as we release it, the movable magnet will spin around so its S pole will be near the N pole of the stationary magnet. That's an electric motor—almost.

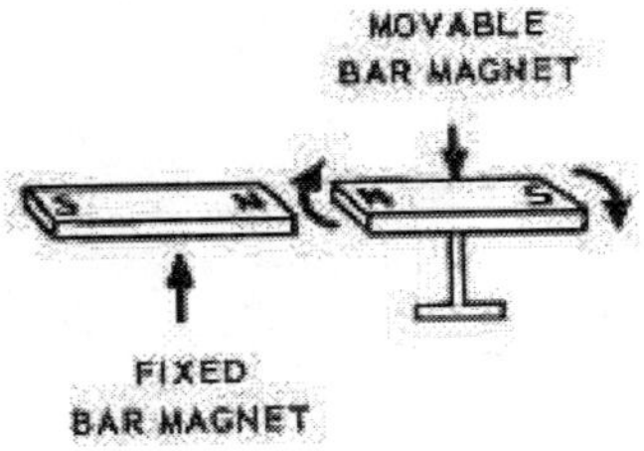

Figure 3. A movable bar magnet pivots so its South pole is near the North pole of a stationary magnet.

It's not quite a motor because the rotating magnet will just move as far as it has to in order to get the opposite poles together. You might be able to cause the movable bar magnet to make turn after turn. You could do this by turning the fixed magnet quickly end for end. This wouldn't be very practical as a motor.

We Can Improve It

If we could change the pole on one end of the rotating magnet just as soon as it reaches the attracting pole, it could make a complete circle. In doing that, the pole at the near end of the rotating magnet would be repelled by the stationary magnet and pushed away. As soon as the opposite end of the rotating magnet would come into the magnetic field, it would be drawn to the stationary magnet. In order to keep the "motor" running, we would have to constantly change the poles at each end on every half revolution.

We Need An Electromagnet

We can't reverse the poles on simple bar magnets, but we can on *electromagnets*. We can make one by wrapping a wire several times around an iron core to form a coil. This magnet will also have a N and a S pole when connected to electrical current. The big difference is that the poles can be changed instantly by reversing the current in the wire.

Switching Poles Automatically

The rotating electromagnet will have to be connected to the 2 wires through which we pass the current. Since it's rotating on a center shaft, we can't have a solid connection. Instead we have to extend the wires from the coil out along the shaft and let the electric contact be made with brushes which touch the wires along the shaft.

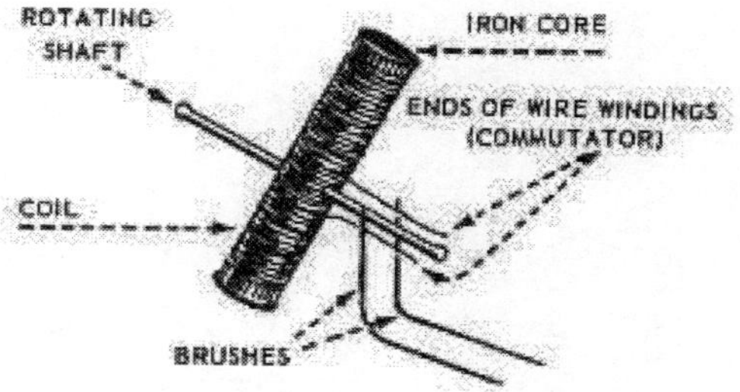

Figure 4. A rotating electromagnet changes poles as contacts are made first one way, then the other.

This is a simple way to reverse the current in the coil of the electromagnet.

Increasing Efficiency

Instead of using only one pole of a stationary magnet, we can use both. This is done by shaping the stationary magnet around the path of the rotating electromagnet. This way we have the benefit of the attracting and repelling forces from both poles. The effect is doubled.

We can also wrap wires around this circular iron and make an electromagnet of it. But when we wire this magnet we use no brushes because we want the current to flow in one direction only.

The stationary electromagnet is called the *field*. The rotating electromagnet is the *armature*.

WHAT TO DO: Make A Motor

Tools Needed:

Pocket knife, hammer, vise (or 2 pairs of pliers).

Materials Needed:

1 roll of No. 24 enameled wire

1 roll of electrician's tape

3 - 4" (20-penny) nails

4 - 2-1/2" (8-penny) nails

4 - 3" brads (10 penny)

Wood board for motor base

2 staples or 4 small brads

2 tacks

2 - 3 volt dry cell batteries (or a 6 volt transformer).

Step No. 1-Armature

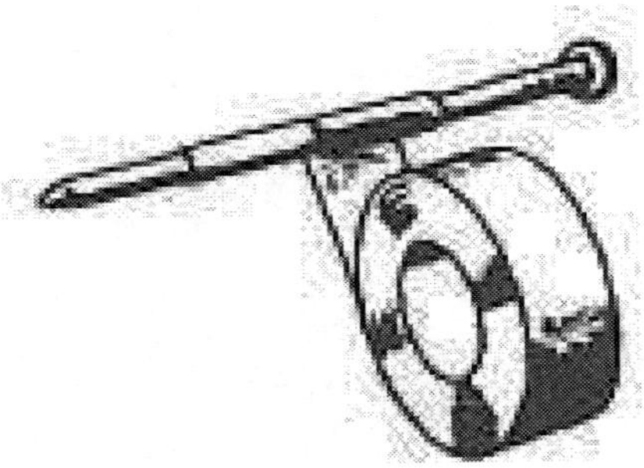

Wrap about 1-1/2" of a 4" nail with two layers of tape. This will be the shaft.

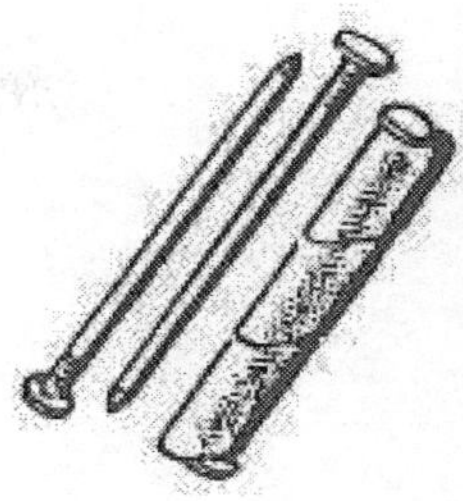

The iron core will be made of two pairs of 2-1/2" nails. Wrap tape around each pair with heads and points alternated.

Center both pairs on each side of the shaft. Place them about 1" from the head of the shaft nail. Wrap them together with two layers of tape from tip to tip.

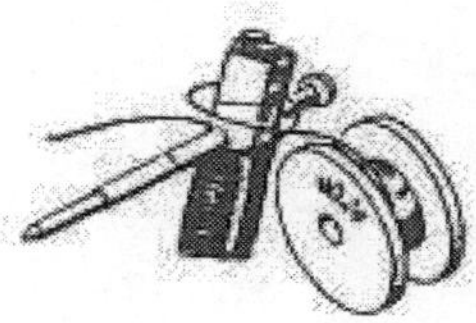

Start at the shaft and wind No. 24 enameled wire to one end and back. Then do the same on the other end. Always wind in the same direction. Leave 6" of spare wire at start and finish.

Step No. 2-Commutator

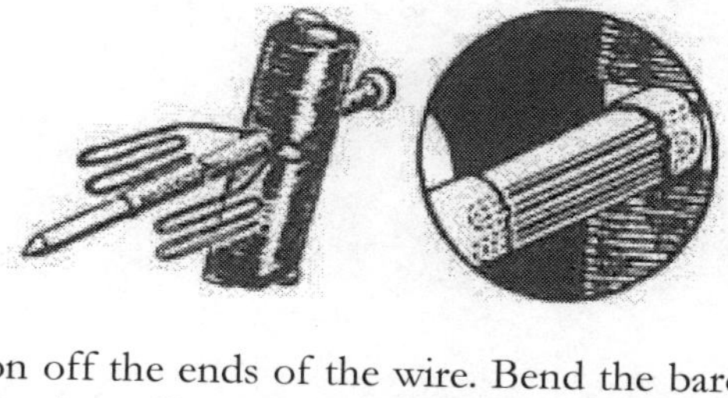

Scrape all insulation off the ends of the wire. Bend the bare ends back and forth as shown. Lay them flat over the taped shaft-one on each side of the shaft.

Hold the commutator down with narrow strips of tape. Wrap tightly near the core and at the opposite end.

Step No. 3-Field

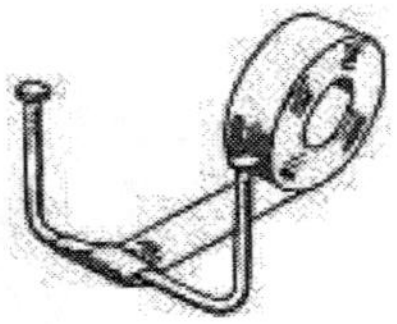

Make the core by bending two 4" nails in the middle at right angles. Space the heads about 3" apart to form a horseshoe. Wrap together with two layers of tape.

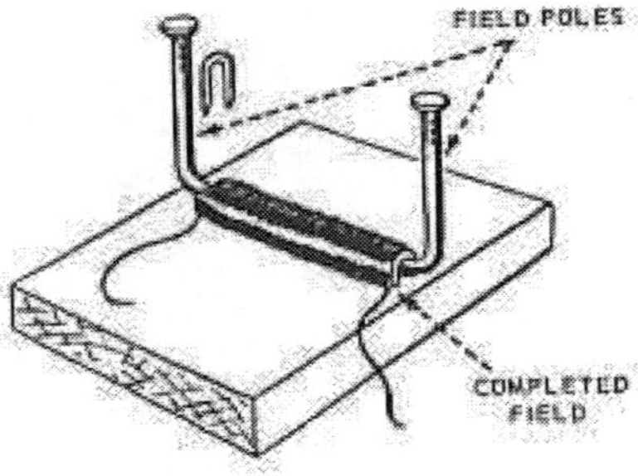

Wind about 400 turns of wire around the center. Leave 4" of spare wire at start and finish. Attach to wood base with staples at each end of the wire. Small brads, bent over, will do just as well.

Step No. 4—Armature Supports and Brushes

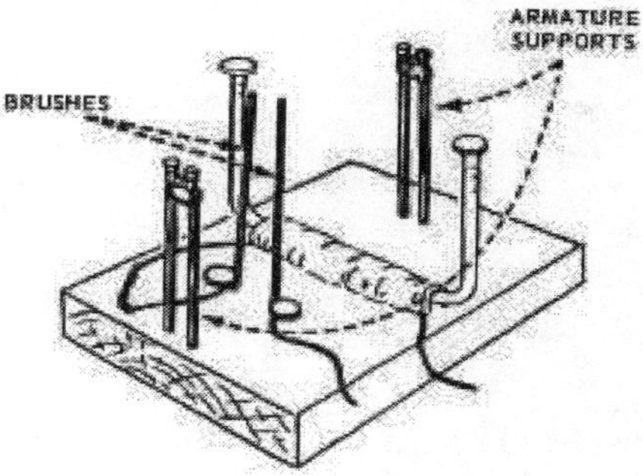

Scrape the insulation from the ends of two 6" pieces of wire. Tack them to the base and bend them as shown to make brushes.

Drive two pairs of 3" brads into the base about 3-1/4" apart and in a line midway between the field poles. Wrap wire around the supports to form armature bearings.

Scrape insulation off ends of wire from the field. Connect one end to a brush wire.

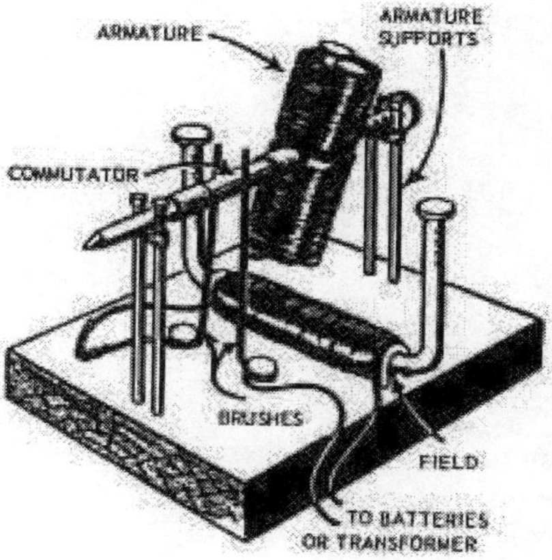

Assemble As Shown

Adjust the position of commutator and tension of brushes against it for best operation.

Take the armature off the motor and connect the commutator wires to a dry cell battery. Test the polarity of each end of the armature with a compass. Switch the connections on the commutator and test again. See how the compass needle changes direction?

With the armature still off, connect the field coil directly to the dry cell. Test the polarity of each end of the field with the compass. How can you reverse the polarity? Try it. It's easy.

Reassemble the motor again and start it. Push the field poles slightly out of alignment with the turning armature. What happens to the motor's speed? Can you tell why?

This time, push the field poles completely out of the way. Test the polarity of the armature as you slowly turn it by hand. Do you see what happens and why it does?

Try to reverse the direction of rotation of your motor by reversing the connections at the battery. What happens? Can you explain why?

Demonstrations You Can Give

Make a display board showing the parts of the toy motor and explain how each part works compared with the parts of a commercial motor.

For Further Information

There are several other types of toy motors you can build. Your club leader or power supplier can help you find information about them.

1. Did your toy motor run?

2. Did your motor speed up or slow down when you pushed the field poles out of line?

Why?

3. What happens to the magnetic polarity of the armature when you turn it slowly by hand and check it with a compass?

4. How can you reverse the direction of rotation of your toy motor?

Is there another way too?

What is it?

LESSON NO. B-6

Credit Points 3

TAKING CARE OF ELECTRIC MOTORS

Through the magic of electric motors, much of our work is done faster and better at lower cost than we could do it without the help of the electric motor. People who use motors and treat them properly have much more time for other work and for leisure time activities. A 1/4-horsepower motor running quietly and steadily hour after hour will do the work of one man, and operate all day for about 5 cents without tiring. On many jobs it will work without "supervision", turning on and off automatically, as required. It does this on water pumps, in heating and cooling units, and on fans and similar appliances.

All that a motor needs to do its work is electricity and a little care. Let's see what you can do to give proper care to motors in your home and on your farm.

You'll Need

A light oil (SAE 10) for motors of less than one horsepower and a slightly heavier oil (SAE 20) for larger motors. See if you need grease for cups which may be on large motors. If so, be sure you use ball-bearing grease and not ordinary cup grease. Cotton waste or clean rags will be needed for wiping off the motors, and a tire pump or vacuum cleaner for blowing out the dust or dirt.

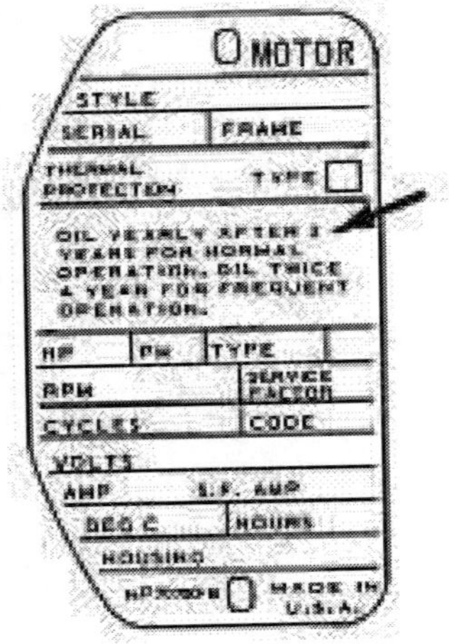

Some motors have instructions for oiling on the name-plate.

WHAT TO DO

1. First, make a list of all the electric motors that work for your home. You may wish to make a separate list for your farm buildings. You'll probably be surprised at how many there are. Don't forget the sewing machine, the refrigerator, the freezer, the vacuum cleaner and other small but important motors. Don't touch any motor that is running. Disconnect them before you touch them.

2. Make a motor service chart with columns headed: Use, Location, Horsepower, Volts, Amperes, Service Required, Date Serviced and What was Done. (See sample) Then list all the motors that require any servicing. Some will have the instructions on the motor or appliance; the instruction booklet that came with the motor or appliance will also tell what servicing is required.

Step 1. Plan the job. Start with the motors in the home. Then you can care for the motors on the farm.

Step 2. Be sure that any motor on which you are going to work is disconnected. Then wipe the outside case clean with a cloth. If the motor has openings in the end, use a vacuum cleaner to suck out dust, dirt or chaff. A tire pump may also be used to blow out this dirt. If you use compressed air, be sure the pressure is not high as it may damage wiring inside the motor. Dust-proof motors should be used in dusty or dirty places.

Step 3.

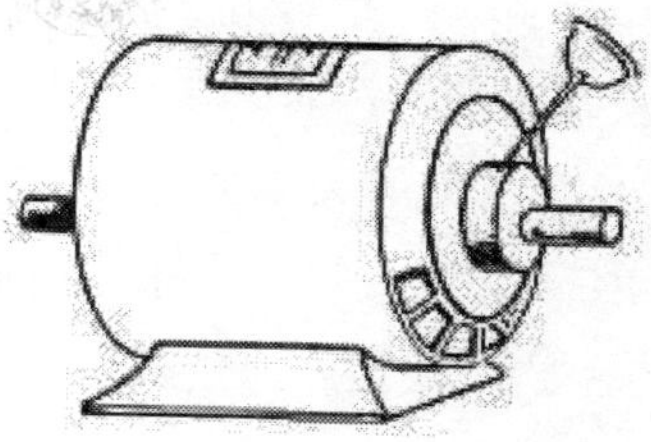

If there are oil holes, oil according to the manufacturer's instructions.

If there are no instructions, remember a little oil goes a long way as far as motors are concerned. Motors of less than one horsepower require only 3 or 4 drops (not squirts) of oil every 3 or 4 months if the motor is used frequently. Too much oil can damage the motor. It spoils the insulation.

If there are no oil holes or grease cups on the motor, it is probably lubricated by means of grease sealed in the bearings at the factory, or it may use greaseless bearings, and does not need to be oiled or greased periodically. Indicate on your chart all motors which need periodic care and see that it is given according to schedule.

Wipe away any excess oil or grease. Be sure oil holes are capped or covered.

Step 4. Reconnect motor and run for a moment.

Step 5. Record on the chart the date you serviced the motor and what was done.

What Did You Learn?

How many motors are there in your home? ______ On the farm? ______

How many motors need regular oiling or grease? ______

How many are less than one-horsepower? ______

SAE Oil ______ is used to oil motors up to 1/2 horsepower. How much oil?______

SAE Oil______ is used for larger motors.

Demonstrations You Can Give

1. Show how to clean a small motor.

2. Explain proper lubrication of motors.

3. Using the chart prepared in this work sheet, give a talk about the motors that work for you-the job each one does, which ones need oil or grease, which need no attention, and why, etc.

4. Use a homemade toy motor to explain "what makes motors run."

5. Show proper way to replace worn cord on a small motor.

For Further Information

Ask your county Extension agent or 4-H leader for more literature on motors. They can help you obtain a film or a speaker such as a power supplier, a local electric dealer, or electrical contractor to discuss motors.

Also visit your public library and see a science teacher for more information on motors.

ELECTRIC MOTORS SERVICE CHART Sample

Use a table like the following to list the motors around your farm and home.

Motor Use	Location	H.P.	Volts	Amps.	Service Needed	Date Serviced and what was done
Food Mixer	Kitchen	1/6	120	4.4	Clean & Oil; cord needs repair	9/1-Cleaned w/cloth. Oiled w/#10 Oil; repaired cord
Tool Grinder	Farm Shop	1/4	120	5.8	Clear, oiling; Have switch repaired	10/6-Cleaned w/vacuum; Oiled, #10 oil. 10/20. Had switch repaired
Pump	Pump house	1/3	120	7.2	Oiling, cleaning	9/26-Cleaned w/tire pump; oiled w/10 oil

LESSON NO. B-7

Credit Points 4

READING THE ELECTRIC METER

There is no question but what electricity is one of the lowest cost services in the home and on the farm. A few pennies worth of electricity will provide the power to run machines that take the place of a man or of several men working all day. However, we all like to know what things cost.

Sometime you may have to decide between different methods—man, horse, gasoline engine or electric motor power. Then you'll want to know how to figure the cost of electricity, as well as the cost of the original equipment. First of all, you should know how to read an electric meter.

Reading a Meter

Electric meters read in kilowatt hours, just as a water meter reads in gallons and a gas meter in cubic feet. A kilowatt hour is the electrical energy consumed by 1000 watts of electricity used for one hour. Ten 100-watt light bulbs burning for one hour would use one kilowatt-hour—one kwh.

Figure 1. Some meters give the reading directly, like the mileage total on a speedometer.

Some meters are read directly, as shown in Figure 1. The more common type has four dials which are read from right to left—just the opposite from the way things are usually read. The hand on the extreme right turns clockwise, the next hand turns counter-clockwise, the next clockwise; the last hand on the left turns counter-clockwise.

The first dial on the right can register up to 10 kilowatt-hours; the second up to 100 kwh; the third, to 1000 kwh; the fourth, to 10,000 kwh. After that, the meter starts over again. To take a reading you must read all four dials of the meter, from right to left.

Figure 2. Meter dials are read from right to left.

To read each dial, you use the number last passed by the dial hand. This may not be nearest the hand. For instance, if the pointer has passed 6 and is almost on 7, you read it as 6. Write down the figures in the same order you read the dial, from right to left. Practice reading the meters shown in Figure 3 on the following page.

What's Your Electric Bill?

Meters aren't set back each month when the meter reader comes around. The difference in the readings from one month to the next shows how many kilowatt-hours have been used. If you know your electric rates, you can figure your bill by yourself. Your power supplier will furnish you with a rate schedule on request.

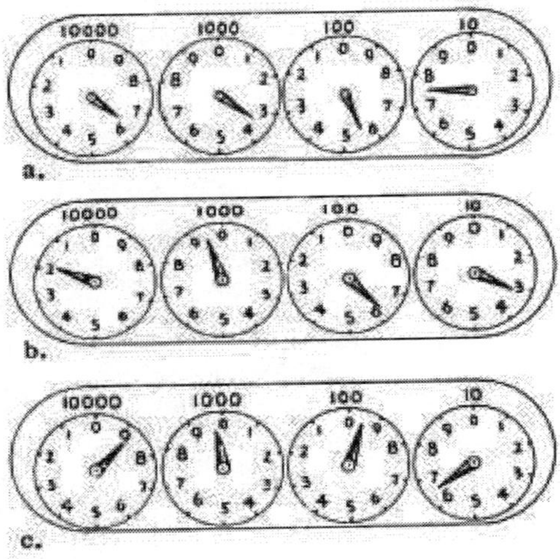

Figure 3. See if you can read the above correctly. The answers are shown in a box on the next page.

It will be interesting to you to find out how much it costs to operate the various electric appliances in your home. A sample rate schedule is shown in Figure 4.

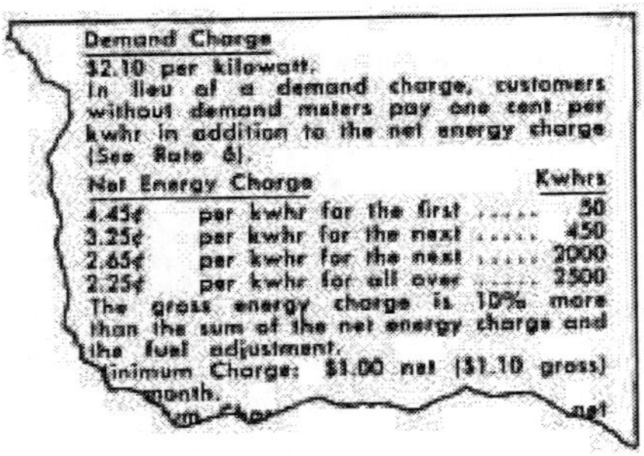

Demand Charge
$2.10 per kilowatt.
In lieu of a demand charge, customers without demand meters pay one cent per kwhr in addition to the net energy charge (See Rate 6).

Net Energy Charge		Kwhrs
4.45¢	per kwhr for the first	50
3.25¢	per kwhr for the next	450
2.65¢	per kwhr for the next	2000
2.25¢	per kwhr for all over	2500

The gross energy charge is 10% more than the sum of the net energy charge and the fuel adjustment.
inimum Charge: $1.00 net ($1.10 gross)
month.

Figure 4. Sample rate schedule. Note that as the use of electricity increases, the average cost per kwh is reduced.

Estimating Operating Costs

To find the cost of operating any single appliance, three steps are necessary:

1. Learn the wattage of the appliance.

2. Estimate how many hours the appliance is used.

3. Find its operating cost.

To Find Wattage:

Watts, you know, are the measure of electrical power. They are the product of voltage (pressure) times amperes (rate of flow). Volts times Amps equals Watts. The nameplate on the appliance will give the voltage required for proper operation as well as either amperage or watts. If it gives wattage, you have the information you want. Otherwise you must multiply volts times amps to get the wattage. When voltage is given as 110-120, use 120 as your voltage. 120 volts is nominal today.

How Much Will You Use?

Now that you know the wattage of the appliance, multiply this figure by number of hours the equipment operates in one day. Divide this by 1000 to get the kwh. Now multiply the result by the number of days the appliance is used each month. This tells you the number of kwh used by the appliance during the month.

Example No. 1

Yard Light: 300-watt lamp

Amount of use: 3 hours per night.

Multiply lamp wattage times hours of use per night to get watt-hours per night.

300 times 3 = 900 watt-hours per night.

Divide watt-hours by 1000 to get kwh per night.

900 divided by 1000 = .9 kwh per night.

Multiply kwh per night times 30 to get kwh per month.

.9 times 30 = 27 kwh per month.

If the yard light is used 3 hours per night, it consumes 27 kwh per month.

Example No. 2

Coffee Maker: 120 volts, 550 watts (from nameplate)

Amount of use: 1/2 hour per day.

Multiply wattage of coffee maker times hours of use per day to get watt-hours per day.

550 times 1/2 hour = 275 watt-hours per day.

Divide watt-hours by 1000 to get kwh per day.

275 divided by 1000 = .275 kwh per day.

Multiply kwh per day times 30 to get kwh per month.

.275 times 30 = 7.250 kwh per month.

If the coffee maker is used l/2 hour daily, it consumes 7.25 kwh per month.

Calculate Operating Cost Per Month

Now that you know the number of kilowatt hours an appliance uses, go to your rate schedule and your electric bill to see what the average kwh costs. Find the average cost of 1 kwh by dividing the amount of your bill by the total number of kwh used in a month.

Example: 410 kwh used.
$14.35 total monthly bill

Average cost per kwh equals $14.35 divided by 410 kwh-3-1/2 cents per kwh.

Therefore, the cost of operating the coffee maker for a month would be 3-1/2 cents times 7.25 kwh—25.4 or 25 cents. Cost of operating the yard light would have been 94.5 or 95 cents a month.

(a) 6357 nbsp; (b) 1963 (c) 8996

Correct answers to the meter readings shown on the preceding page.

Adding Low Cost Helpers

You can see, by looking at your rate schedule, that the average cost per kwh gets lower as you use more electricity. To find the cost of operating additional electrical equipment, the cost per kilowatt hour is found from the last "step" in the bill—the lowest cost per kwh of the electricity you're now using.

Sometimes power suppliers give special rates for such equipment as electric water heaters.

WHAT TO DO: Find the Cost of Operating Electrical Equipment

Make and fill in the blanks of a chart showing the electrical equipment you have and the operating costs per month.

Make a chart for the home (refer to chart one). Show the probable operating cost of equipment you might add to what you now have.

Demonstrations You Can Give

Show how to read a meter, making one with plywood or cardboard. Dials can be painted on the main board. Arrows can be attached so they will revolve to give different readings.

Show how to find the wattage of various types of equipment.

Show how to figure the cost of the average kwh in a home.

For Further Information

Your leader can get additional material for you or you may want to have someone from your power supplier talk to your club, telling about meters, how they work and how they are regularly checked for accuracy.

Chart One-THE HOME

Column No.	1.	2.	3.	4.	
Item	Wattage Rating	Hours Used per Month	KWH per Month (col. 1 x 2) / (1000)	Cost per Month (Col. 3 x av.kwh cost)	Remarks
Electric Iron	1100	30	33	.80	
Stove	880	60	52.8	1.21	(Special rate.)

LESSON NO. B-8

Credit Points 3

IRONING IS FUN WITH THE MODERN HAND IRON

When you are getting ready to go to school or to a party, it probably gives you a good feeling to put on a clean, freshly-ironed skirt, blouse or dress. But did you ever think about the electric iron that helps so much to give you that well-dressed feeling? When you were younger, you may have had a play iron and pretended to iron your doll's dresses. Now you are old enough to learn about real irons—the different kinds of irons, how the iron heats, the kind of cord needed, the type of outlet necessary, how to use safety rules when you iron, and even how to help with the ironing.

Important Things to Know

There are many different irons, but the two kinds most important for you to know about now are the regular dry iron and the combination steam-and-dry iron.

The thermostat keeps the iron at an even temperature.

It isn't weight alone that makes an iron do its job, but the heat of the iron. The heat is given off in the sole plate. The automatic iron has what is called a *thermostatic* control which holds the temperature of the iron at the heat you want. Some clothes need to be ironed with a very hot iron, while others need only to be pressed lightly with a cool iron. The thermostat keeps the iron at an even temperature after you set it for the heat you want. The thermostat is the heart of the iron.

Take a look at the iron used in your home. It isn't heavy to lift, and has a handle that fits your hand easily. It looks graceful and has a smooth bottom, called the sole plate. And it may have a narrow, pointed tip which is helpful in ironing pleats, corners and gathers.

Your iron has a smooth bottom called the sole plate.

The Iron and Safety

If you are going to learn to do some ironing yourself, the most important thing for you to remember is SAFETY. You should read all about the iron first in the instructions which came with it.

Never use an iron carelessly. Remember the safety rules:

1. An iron should never be left even for a few minutes without being disconnected. Turn off by removing the plug from the outlet, or by turning the control lever to "off."

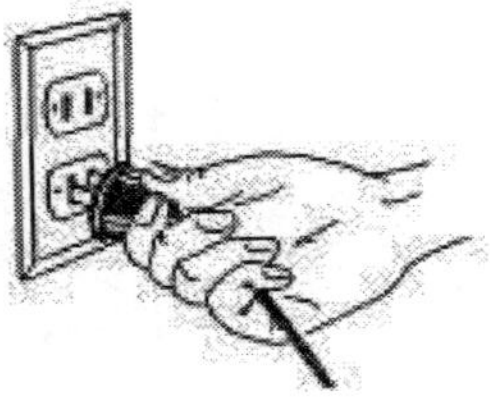

Take hold of the plug—not the cord—when you disconnect it from the outlet.

2. Let the iron cool before putting it away.

3. Wrap the cord carefully around the iron after it is cold.

4. Always stand the iron where it will not fall off on a child or pet or your own toes.

WHAT TO DO: Learn About Your Iron

Materials Needed: An automatic iron, some old play clothes, towels, napkins or handkerchiefs, and an ironing board.

Steps to Take:

1. Watch an experienced person iron.

2. Ask questions about what clothes need to be sprinkled.

3. Study the thermostat settings on the dial or indicator.

Most irons have a dial to set for the proper heat for different fabrics.

4. Ask about the kind of fabric each piece of clothing is—cotton, linen, silk, nylon, etc.—and why the iron should be at high heat for some, cooler for others.

5. Set the thermostat for the amount of heat needed, and with an older person watching you, iron some handkerchiefs, napkins, bath towels, and a pair of play shorts or blue jeans.

6. During a month iron some of these articles for your family, keeping a record of how many you do and what they were.

7. Take care of your iron. Be responsible for storing it.

Date	No. Aricles Ironed	Type of Article	Stored Iron Properly (check)

IRONING IS FUN

1. I (use) (do not use) an adjustable ironing Board at home. If I do, I adjust it to the height that just clears my knees easily as I sit in a comfortable chair. Yes No

2. There are three kinds of irons usually used—dry iron, steam iron or a combination steam or dry iron. I use a —— iron.

3. I (have) (do not have) the instruction book. (If you do, read about the iron.) I know the iron's parts by their correct names. They are——.

4. I disconnect the iron if I leave it even for only a minute. This is a safety measure as fires have been known to start from irons left connected. Yes No

5. I take hold of the plug—not the cord—when disconnecting the iron. Yes No

6. I wait until the iron is cold before wrapping the cord around the handle and storing the iron because——.

7. Most irons have a thermostatic control. The iron I am using has settings for——.

8. The purpose of the thermostat is——.

9. These fabrics need high temperature.——

These fabrics need medium temperature.——

These fabrics need low temperature.——

10. These fabrics need sprinkling.——

11. The heat and smoothness of the sole plate smoothes the wrinkles. Pushing down on the handle or moving the iron rapidly only makes ironing hard work. I will iron slowly and steadily arranging and moving the garment with the left hand while guiding the iron with the right hand. (Or the other way for the left handed.) Yes No

12. I have watched an experienced person iron. Yes No

13. I have practiced on handkerchiefs, napkins and pillow cases.

14. Here is my record of ironing for one month

Month____________

Your Name ______________________

Date	I have ironed:

Demonstrations You Can Give

1. Show a dry iron and a steam-and-dry iron. Tell the difference between them and when each is to be used.

2. Display garments that look nice because they have been ironed properly, and those that have been ironed improperly. Explain about the heat, thermostat, type of iron and why results differ.

For More Information

At a club meeting ask a parent to give a demonstration of ironing different articles. Some power suppliers or dealers have people who will demonstrate the proper way to iron, and how to care for irons.

LESSON NO. B-9

Credit Points 2

LET'S BE FRIENDS WITH ELECTRICITY Plan a Hazard Hunt

Electricity can be your important lifelong friend and helper, so you will want to know all you can about it and how to treat it properly. However, careless and improper use of electricity can do a lot of harm. Used properly, and treated with respect, electricity can do wonderful things to help you every day in many ways.

For safe and proper use of electricity, all wiring, fittings, insulation, cords and plugs must be in good condition. You can be a detective and track down defects in any such type of electrical equipment that you may be using in your home or on your farm.

When you find anything that is wrong, and know where it is, and know what to do about it, you can very likely correct the condition yourself, such as replacing a worn extension cord with a new one. If you find defects in permanent wiring, or some places where wires are bare or terminals are needed, you should tell your parents about them.

SAFETY FIRST, remember, should always be on your mind when working with anything electrical.

WHAT TO DO:

1. Have A Hazard Hunt

Go on a Hazard Hunt to see how many electrical hazards you can find. Look for defects such as broken insulation, worn cords, splices that are not properly soldered and taped, loose connections, or switches that aren't working properly.

There are many ways to have a Hazard Hunt. Choose the method that will be the most fun. Use the Hazard Hunt Guide in this outline to check your home, and other buildings. Maybe you'll want to have a friend help check your home, then you help him check his. Or, why not give each member of your family a Hazard Hunt Guide and have a contest? Parents may want to team up against you and other younger members of your family to see which team can find the most electrical hazards in some set time—say 30 minutes.

Have a Hazard Hunt Committee in your club check all member's homes and buildings and report its findings at the next club meeting.

To Make It More Fun

1. Put a hazard tag, like the one shown, (Figure 1) by each hazard that is found. Leave it until the hazard is corrected. Have another contest to see which member of the family corrects the most hazards.

Figure 1

2. Report on your Hazard Hunt at the next club meeting. Tell about the Hazards found, and what you have done or plan to do about them.

3. Suggest that the entire club have an Electric Hazard Hunt at your club meeting places or any community building. This could be part of one meeting.

4. Have a contest between two teams in the club to see which team can get the most homes in your community checked by the Hazard Hunt Guide. Losers could give a party for the winners.

2. Get Others Interested

Promote a community Electric Hazard Hunt. Enlist the support of power suppliers, electric supply and equipment dealers, schools, newspapers, radio and television stations.

What To Look For

Make a complete tour of your home and other buildings and see how many hazards you can locate. When you find a hazard, put a tag near it to mark it.

SAFETY TIPS

Put hazard tags *near* the hazard but *not* directly on broken or frayed wires, insulators, fittings, or other wiring equipment. Do not touch them either. Badly-frayed wires should be disconnected immediately from the power supply. In this way, you will not expose yourself to shock by accidentally touching an exposed live wire that may be carrying current.

4-H Electric Hazard Hunt Guide

Wiring and Protective Devices

1. Cable or conduit splices not in boxes——

2. Cable or conduit not securely clamped in boxes——

3. Conduit or armored cable not properly grounded——

4. Cracked or broken insulators (Figure 2)——

5. Wire not completely covered with insulation——

6. Worn insulation on wire——

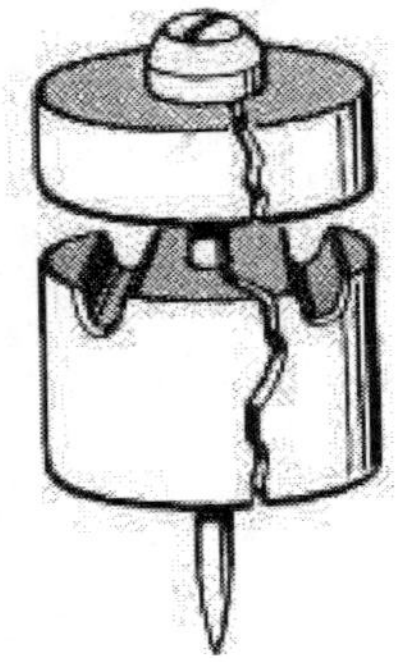

Figure 2

7. Old unused wiring not yet removed——

8. Outlets, junction and switch boxes not securely fastened and covers not in place——

9. Switches not working properly (sparks fly as switch is flipped) (Figure 3)——

10. Fuses not of proper ampere rating for circuit——

11. Extension cord used in place of permanent wiring——

12. Pull chain socket without an insulating link in the chain——

13. Pull chain socket near plumbing fixtures or where hands may be wet or one may stand in water——

Figure 3

14. No moisture-proof cords for outside weather conditions or heavy rubber cords for motors and motor driven appliances

Lighting

1. Fixtures in farm buildings installed so that they might be easily damaged

2. Lights in haymows and other dusty locations not protected by dustproof globes

3. Outside sockets not waterproof

4. Heat lamps not properly supported by non-current carrying wire, chains, or brackets (Figure 4)

5. Light bulbs not frosted, shaded, or placed so that light is diffused to prevent glare

Figure 4

Auxiliary Wiring

1. Outlets overloaded—in other words, "octopus wiring"
2. Extension cords placed under rugs
3. Extension cords run through doorways (Figure 5)

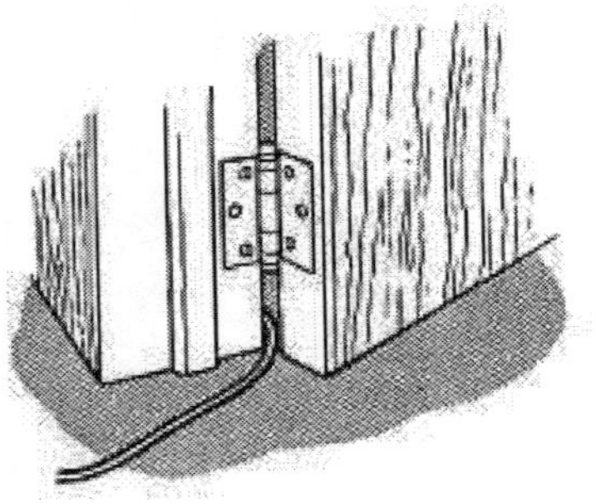

Figure 5

4. Extension cords or lamp cords should use underwriters' knot (Figure 6)

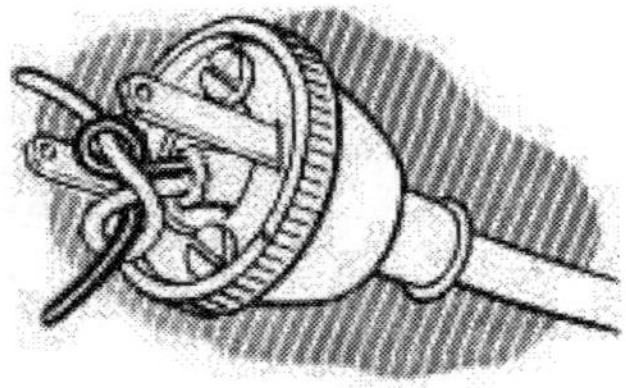

Figure 6

5. Plug connections fuzzy (Figure 7)

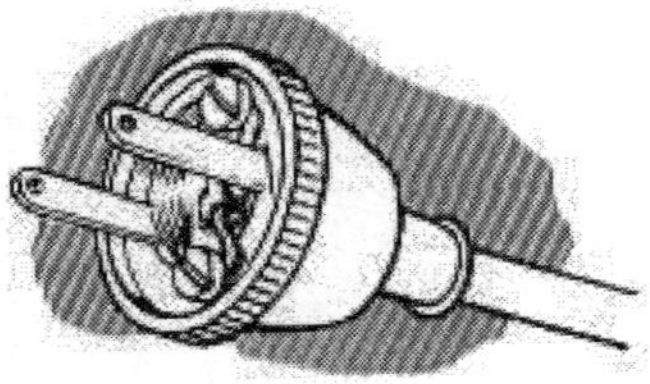

Figure 7

6. Extension cords run over heaters or radiators
7. Extension cords, or appliance or lamp cords, worn or frayed
8. Heating appliances without regular asbestos covered wire

9. Open sockets or outlets where a baby or small child might stick a finger or metal toy

10. Broken plugs (Figure 8)——

11. Loose prongs on appliance or lamps plugs——

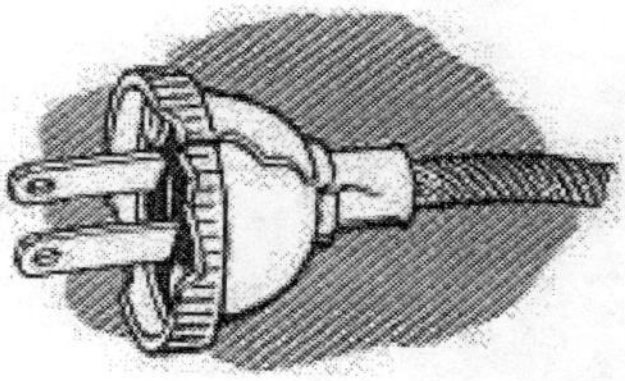

Figure 8

How Many Hazards Did You Find?

Make a chart listing the hazards, their locations and what you did about them. Make your own chart and list what you find.

Demonstrations You Can Give

Show and tell others how to have a Hazard Hunt.

For Further Information

Check with your leader, then ask your power supplier or a local electrician to tell you about safe electrical wiring, connections and fixtures.

Hazard	Location	What I Did
Loose prong on lamp plug	*Living Room*	*Replaced with new plug*
Cracked insultor on service wire in house	*Back of house*	*Notified power supplier*
Conduit not securely clamped to box	*Basement by fuse box*	*Notified parents*
Extension cord, old and worn	*Basement, by washing machine*	*Replaced with new rubber-covered one and protected it from water*

LESSON NO. B-10

Credit Points 3

HOW ELECTRIC BELLS WORK—FOR YOU

When was the last time you wanted to get a simple message like "You're wanted on the telephone," "There's someone here to see you, "or "There's a car in the driveway," to someone around your place? Did you have to walk or run some distance and perhaps shout, too, to be heard by the other person? Perhaps you had to stop some other work, or interrupt your favorite kind of fun, to do this bit of messenger work.

If the nature of the message is like one of those mentioned, and the number of people in hearing is not too great, then perhaps you can use bells or buzzers or both to do some of your messenger work for you. Even though a bell or a buzzer can't talk, it can convey a message.

What to Do

1. Learn how bells and buzzers work, and learn about the many different kinds.

2. Plan and install a bell system for your home or farm.

Bells and Buzzers Can Tell a Lot

Electric bells and buzzers use the same basic principle as the telegraph system, invented by Samuel Morse in 1840. Although not as important today as it was before radio, telephone, and teletype became common, the telegraph is still in use.

Bells and buzzers, however, are very common and have many uses. They are most often seen in the form of doorbells, and rare is the new home that does not have one or more. Service stations have bell systems to let the operator know that a car is waiting at the gas pumps. A clock signal reminds the homemaker when the cooking time is completed. Children are called to and released from school classes by means of bells and buzzers.

Also, various alarms employing bells and buzzers warn us when it's time to get up, or even that the place is on fire, or that a burglar is trying to break in!

Let's find out how bells and buzzers work, what different kinds there are, the different ways you can control them, and how you can put them to work for you.

You'll find that buzzers and bells can help you with your 4-H projects, and with the proper controls, can be your eyes and voice in a dozen places at once.

Why They Buzz or Ring—Electromagnetism

If we were to look at an electric bell with the cover off, we'd find that it would be very much like Figure 1.

A push on the button, which is just a switch that is normally held "open" or off by means of a spring, sends the current from the battery or transformer through the circuit.

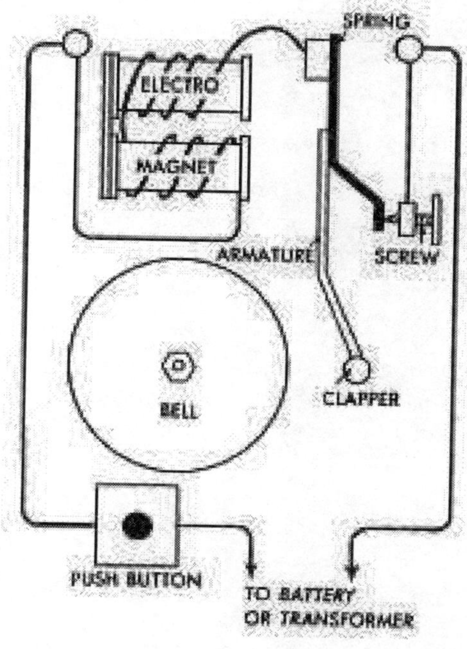

Figure 1

You will see that the current passes first through two small coils of wire, and each coil has at its center a piece of soft iron called the core. When the current is on, the core becomes magnetized and attracts another piece of iron called the armature with its clapper attached.

This action rings the bell, but it also breaks the current by pulling the spring away from the screw on its return to the power supply.

With the power off, the electromagnet lets the spring return the armature to its normal position, contact is made again, and the cycle starts all over again—just as long as you continue to push on the button.

Buzzers work exactly the same way, except that they do not have a bell and depend instead on the vibration of the armature for a noise that's not as loud or as musical.

Gongs or chimes, that strike only once when the button is pushed, are made by connecting the armature with the screw by means of a flexible wire.

A Special Kind of Electricity

Most buzzers and bells work on a much lower voltage than you normally find in the wires in your house. Some are made to work at 6 volts, others at 10 volts, and still others at slightly higher voltages.

You can get these low voltages by using one or more batteries, or by using a transformer connected to your house current. Most bells and buzzers are now powered through transformers.

How to Control Them

The push button is the most common means of control. You can use one button to control several bells, or several buttons to control one bell, or have several buttons control several bells. Because low voltage is used, adding extra buttons is simple, inexpensive, and safe.

Buzzers and bells can also be controlled by: *clocks*, as in the interval timer on an electric range or in a school class bell system; *temperature detectors*, as in a fire alarm or freezer alarm; *door and window trips*, as in a one-man repair shop or in a burglar alarm; and *treadles*, as in the driveway of a service station.

Figure 2

Pick the Right Bell or Buzzer

Some of the many different types of bells, and various ways of controlling them are suggested in the table below. Just remember that no matter what the job or conditions, you can probably find a bell or buzzer and controls that suit your need.

SOME TYPICAL JOBS FOR BELLS & BUZZERS				
Job	Type of bell or buzzer	Number and location of bells and buzzers	Type of control	Number and location of controls
Summon others to the telephone	In the house—small to medium buzzers In outbuildings—medium to large bells Outdoors—large weatherproof bell All transformer-powered	Enough to cover all usual work locations	Push-buttons	One at the telephone and each extension phone
Notify club member that car is at his produce stand	Medium to large bell—transformer-powered	One may be enough—if mounted on the back of the stand	Hose diaphragm (Complete driveway units including control, are available, ready to plug in.)	One—in the driveway
Warn of power failure to incubator or brooder	Battery-powered buzzer, medium size	One near the poultryman's bedroom	Relay, held open as long as power is on, closed by spring if interruption occurs	One, at main switch of hatchery or brooder house
Warn of dangerously warm temperature in freezer	Battery-powered buzzer, medium size	One, in or near the kitchen	Temperature detector (sensitive thermostat)	One, with bulb inside freezer

How to Plan Your System

To save your time and steps when the telephone rings for someone else in your family who is some distance away, you can install a simple bell or buzzer system to summon that person.

First, you must plan what you are going to do. On a large sheet of paper, draw to scale (roughly) a plan of your house and grounds, including those places where phones are located. It will help if you rule off your paper in 1/8" or 1/4" squares and let each square equal one foot. Show the location of poles supporting your wiring.

Next, pick out those areas where you or others would likely be when someone else would answer the phone and want to call you to it.

After you have thought about this, and talked it over with members of your family, show locations on your plan where you think you would like to have buzzers or bells, and show a button beside each telephone. (Generally, you should have a bell or buzzer near each phone, also.)

Figure 3 shows diagrams of various types of systems, and will help you determine the number of wires you will have to install to connect the buttons and bells that you have planned.

Inside, you will connect your transformer and the various buttons and bells with ordinary indoor bell wire. Outdoors, however, you should use weatherproof 2-wire or 3-wire telephone twist.

Show on your plan the distances that must be traversed by each type of wire, and show the number of conductors in each. Don't overlook the vertical distances (one floor to another).

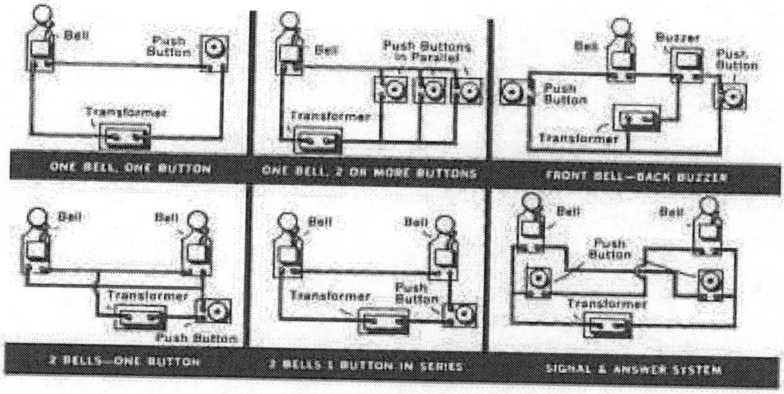

Figure 3

Materials You'll Need

Because no two situations are just alike, it will be necessary for you to make your own list of materials.

As a guide, however, here is a list of typical materials, with the quantities left blank, for you to fill in as your own requirements and measurements dictate.

10-volt transformer
______ Door buzzers
______ Doorbells
______ Weatherproof outdoor type bells
______ ft. indoor bell wire
______ ft. 2-wire weatherproof telephone twist
______ ft. 3-wire weatherproof telephone twist
______ lbs. staples (insulated)
______ entrance insulators (for attaching weatherproof to buildings and poles)

Because your transformer must be wired into your regular house current, you should have some help on this from an electrician or other qualified person. Also, you should get that person to review your plans and materials list before you place an order.

Install According to Your Plan

With the aid of an electrician or other qualified person, install your transformer, and test it.

You may then go ahead and complete your signal system, checking carefully with your plan, and making sure that your installations are both electrically and mechanically secure.

Test your system in all possible ways that it might be used.

Demonstrations You Can Give

Build a demonstration board incorporating a farm or home layout, with pushbuttons or other controls and bells and buzzers appropriately located. Show and tell how the system would save time and energy.

Show and tell how some of these work, and their value: power-off alarm, freezer alarm, fire alarm, driveway alarm.

For More Information

Ask your power supplier or your nearest electrical supply house for catalogs or literature on various types of signal systems, or ask a dealer to show you equipment he has in stock.

LESSON NO. B-11

Credit Points 2

FIRST AID FOR ELECTRICAL INJURIES

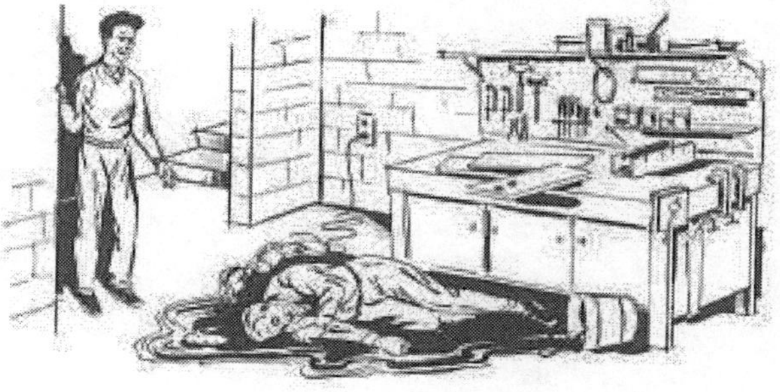

What would you do if you saw someone who had been hurt by electricity?

Did you know that you could save his life, if you had taken the time to learn and practice a few simple rules of electrical first aid?

First aid training equips you to know what to do and what not to do for the injured until medical help can be obtained. While the main benefits are for you and your family, no one can call himself a good citizen if he fails to help a stranger who has been hurt.

The information given here is only for electrical injuries. Perhaps what you learn will inspire you to take a complete course in first aid.

What to Do

Learn how to prevent electrical accidents, and what to do if an electrical accident occurs.

1. Make an electrical hazard hunt in your home or on your farm. Point out to your parents everything that should be repaired or replaced for safety's sake.

2. Read the first aid suggestions that follow. Learn them.

3. Get to know the six steps that are outlined for mouth-to-mouth rescue breathing. Practice them on your brother, sister, or parents. Teach the entire family how to do it.

Electricity Can Kill

In this day of hundreds of uses of electricity, you should know about electrical dangers. Electrocution can occur from either low voltage (household type) or high voltage currents. Sometimes household voltages are more hazardous because people underestimate the dangers involved.

A fraction of an ampere passing through your heart muscles can be fatal. Your body offers some resistance to the flow of electricity to ground. If you are standing on wet ground or in water, or if your skin is damp, this resistance is greatly reduced.

Wire cables within walls and cords on appliances are all insulated with a shock proof covering. Continued use, age, or damage may expose a bare wire and create a hazard. The point of exposure need be only a fraction of an inch. Cords are often used and abused. Exposed wires and signs of wear are danger signals.

Always be wary of overhead wires. People have been injured or killed when kite strings, model plane control lines, irrigation pipe, and water well equipment have come in contact with the power supplier's or their own overhead wiring.

Prevent Accidents

Underwriters' Laboratories (UL) have taken steps to see that minimum safety standards are met in the manufacture of electrical equipment. Look for the UL label when you buy cords or appliances. Never place cords under carpets or furniture, or drape them over a nail. Replace or repair worn cords without delay.

Be especially careful when operating electric devices in the bathroom. Keep in mind the dangers of a wet floor, grounded metal pipes, and wet skin. Turning on an AC radio while you are taking a bath is asking for real trouble.

There may be shorts in electric devices. Keep your hands dry when using them, and do not touch them along with grounded metal objects. If you ever get a slight shock, sound the danger signal and do something about it.

Think, Then Act

Your first thought in rescuing a victim from an electrical accident should be your own safety. Speed is also important, because a few seconds or minutes may save a life.

The first question you should ask yourself is "Can I quickly turn off the power?" This would be easier to do in the home than outside. In the case of a victim trapped in a bathtub from a radio accidentally knocked into the water, it might mean simply removing the plug from the wall outlet. If a victim is found grasping shorted, permanently installed equipment and cannot let go, the main switch might be used for quick release of the current.

Outdoors, especially with high tension wires, your danger in rescue is much greater. To handle the victim, touch him only with a long dry stick, dry rope, or a long length of dry cloth. Be sure your hands are dry and that you are standing on a dry board. A broom might be a good lever to pry a victim from a high tension wire but never use a green stick containing sap.

First Aid

Once the rescue has been made and the victim is free of further danger, check to see if breathing has stopped. If so, start artificial respiration *immediately* and send someone for a doctor.

Artificial respiration must be started as soon as possible after normal breathing ceases. *Most persons will die within 6 minutes or less if breathing stops completely unless they are given artificial respiration.* Precious minutes may have passed before you get to the victim. Since the victim may be within seconds of death by the time you are able to touch his body, you should seek to obtain an air flow to and from the lungs *immediately*.

The victim may seem stiff as an effect of the current, so don't give up easily. Continue the procedure for several hours. If transportation is necessary, remember that there may be internal injury, fractures, or severe burns.

Mouth-To-Mouth Rescue Breathing

There are various effective ways to give artificial respiration, each with its advantages and disadvantages. The mouth-to-mouth method is recommended as a good one to master. It can be used on victims of drowning, suffocation, and asphyxiation, too. People have been known to save lives with less exposure to the correct procedure than you are getting by reading this. So, pay attention and remember what you read.

Step 1. Turn the victim on his back. Wipe out victim's mouth quickly. Turn his head to the side. Use your fingers to get rid of mucus, food, sand, and other matter.

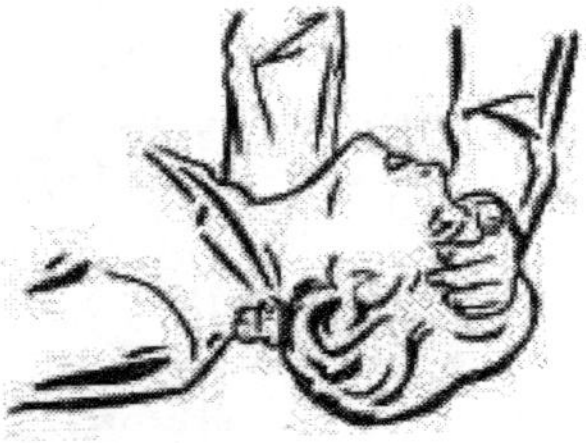

Head Position

Step 2. Straighten victim's head and tilt back so that chin points up. Push or pull his jaw up into jutting out position to keep his tongue from blocking air passage. This position is essential for keeping the air passage open throughout the procedure.

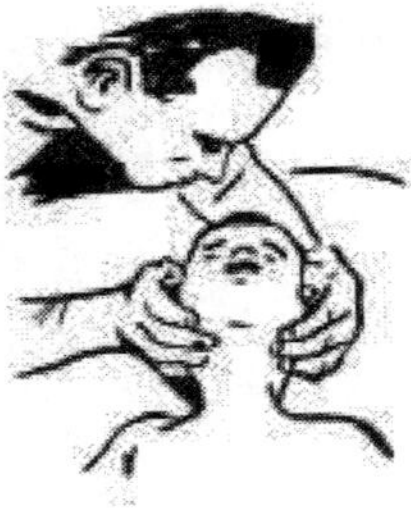

Push Jaw Up

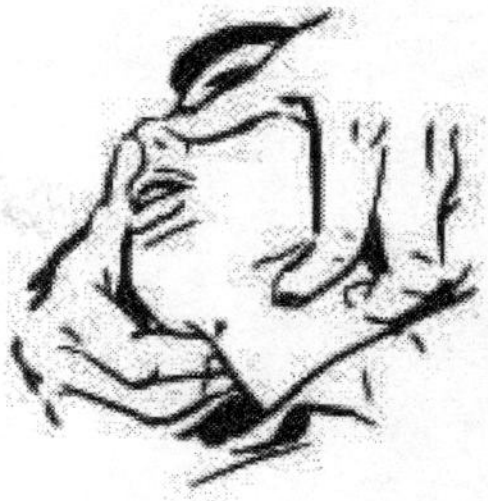

Pinch Nostrils

Step 3. Take a deep breath, place your mouth tightly over victim's mouth, and pinch nostrils closed to prevent air leakage. For a baby, cover both nose and mouth tightly with your mouth. (Breathing through handkerchief or cloth placed over victim's mouth or nose will not greatly affect the exchange of air.)

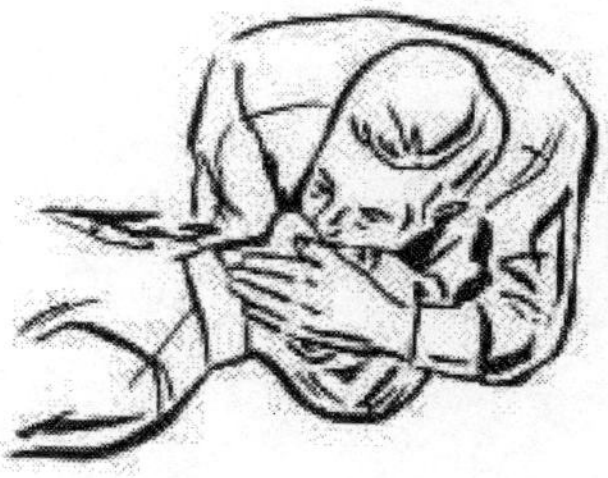

Breathe

Step 4. Breathe into victim's mouth or nose until you see his chest rise. (Air may be blown through victim's teeth, even though they may be clenched.)

Step 5. Remove your mouth and listen for the sound of returning air. If there is no air exchange, recheck jaw and head position. If you still do not get air exchange, turn victim on side and slap him on back between shoulder blades to dislodge matter that may be in throat. Again, wipe his mouth to remove foreign matter.

Step 6. Repeat breathing, removing mouth each time to allow air to escape. For an adult, breathe about 12 times per minute. For a child, take relatively shallow breaths, about 20 per minute. Continue until victim breathes for himself.

What Did You Learn? True or False

1. A broken arm should be splinted before artificial respiration is applied to a victim who is not breathing.

2. A person who has been severely shocked with an electric current should lie down.

3. A doctor should be called even though you successfully have revived a victim's breathing.

4. A fraction of an ampere through the human heart muscles can be fatal.

5. A copper wire would provide a better path than your body for stray currents, therefore all appliances should be grounded if possible.

6. Outside wires are never a hazard because they are covered with insulation when they are installed.

7. Cords need not be repaired until you can see bare wires.

8. Tuning in an AC radio while you are bathing is always dangerous, even though your hands are dry.

9. In an emergency, a broom is an acceptable tool for prying a victim off a high tension wire.

10. In mouth-to-mouth breathing, an adult's lungs should be filled 12 times per minute and a child's 20.

Demonstrations You Can Give

Show how to deal with an electrical first aid "problem" given to you by your leader.

For More Information

Ask your leader to have a first aid expert put on a demonstration. (Many industrial plants and power suppliers have such people.)

LESSON NO. B-12

Credit Points 3

HOW ELECTRICITY HEATS

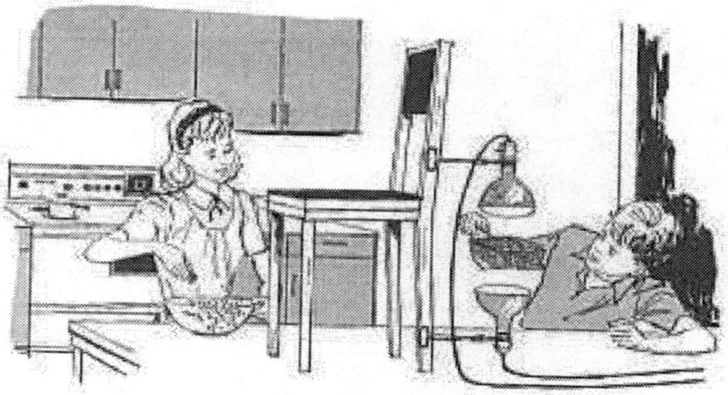

In ancient times, people thought that heat was a material just as air is. They called it "caloric". When something got warm, they said, caloric flowed into it. When something cooled off, caloric flowed out of it. It did not bother them that they could not see caloric. They could not see air either!

Now we know that heat is not a material. It does not take up space. It does not weigh anything. Instead, it is a form of energy. And when we say that heat is a form of energy, we mean that it can be used to do work.

What to Do

1. Make a simple resistance heater.

2. Make some popcorn by:

(a) conduction (b) convection (c) radiation

"Resistance" Makes Heat

There are at least four ways that electricity can make heat. The one that we'll cover here is *resistance* heating. (The others are: *dielectric* heating, where the lines of force of an electrostatic field pass through a non-conductive material and heat it; the *heat pump*, which is a refrigerator in reverse; and *electronic* heating, which uses high frequency waves similar to radio waves to create high speed movement of the molecules or tiny particles which rub together to make heat.)

Resistance heating occurs because every conductor of electricity opposes the flow of current through it. Some conductors resist more than others. When they do, a certain amount of warming takes place. The more resistance that is offered, the more heating there is.

Some materials, like silver, copper, and aluminum, offer little resistance. We say they are good conductors.

Other materials, like iron, offer more resistance. They are still conductors, but not as good as the others mentioned.

The *size* of the conductor, and its *length* are the other two things that affect its resistance. The *smaller* it is, the greater its resistance. Also, the *longer* it is, the greater its resistance. Therefore, when we only want to *move* electricity from place to place, we want relatively large, "good" conductors. Here, we do not want to make heat. In fact, we want to avoid it, because too much heat in the wrong place can cause a fire.

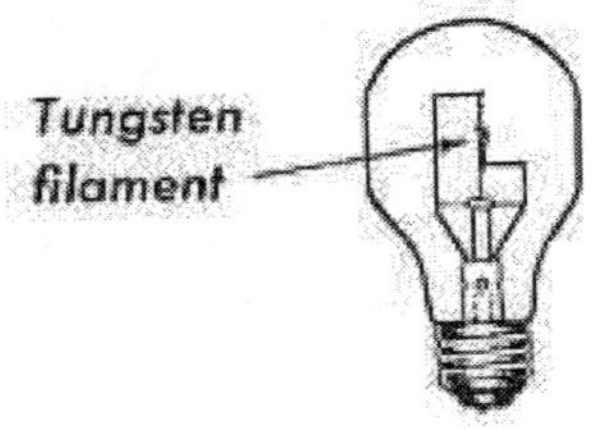

But when we want heat, we choose relatively small, "poor" conductors, and the more heat we want, the longer they must be. If you will think of the filament inside a lamp bulb; you may recall that it is a very fine wire, coiled so as to get a maximum length, and made of tungsten which has a high resistance.

Because of all these factors, this filament glows at a white heat, and is a source of both light and heat.

Make a Simple Resistance Heater

Materials you will need:

1 dry cell battery
1 foot iron picture wire
Pliers

Use a short strand of iron picture wire and hook the ends to the terminals of a dry cell battery. Use pliers so that you do not burn your fingers. Disconnect the wires as soon as they become hot. Tell why the wires heat.

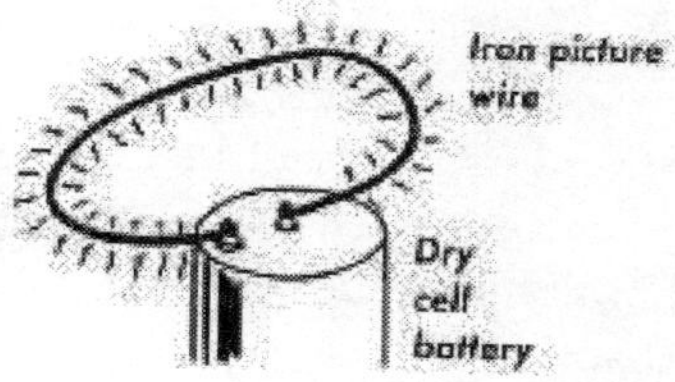

Conduction is "Touching" Heat

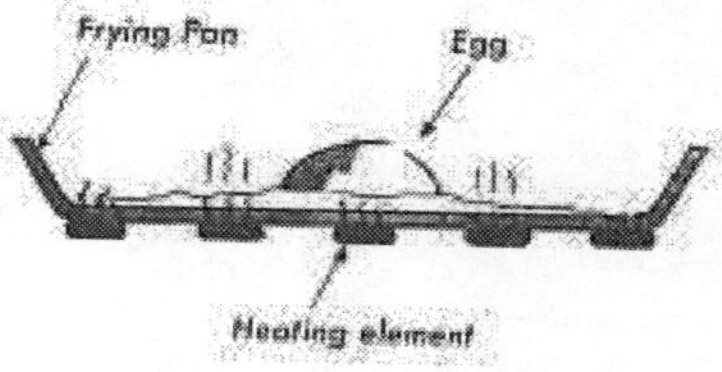

Conduction occurs when you set a pan containing food right on a heating element. An egg cooking in a hot frying pan is a good example of conduction at work. This method is the most efficient single way of using electric heat for cooking.

Convection Depends on Air

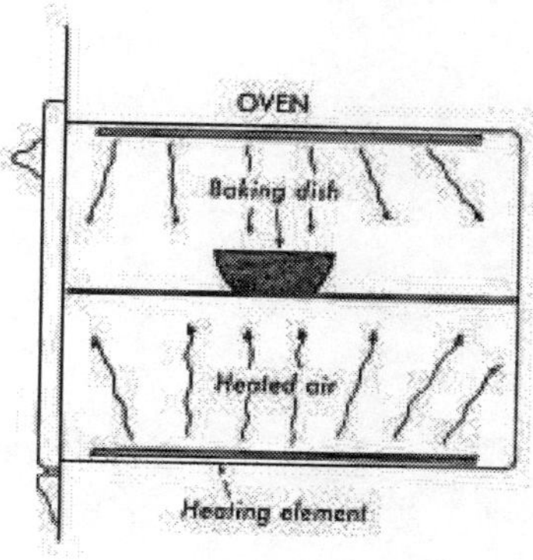

Convection warms food in pans that are not actually touching the heating element. It uses the hot air around the element to carry heat to the pan.

Your oven in your range works by convection. Most houses are warmed in winter in the same way. The heat produced in a furnace warms the air as it circulates through. This air in turn keeps your body warm.

Radiation is Like the Sun

Radiation heating is more difficult to explain. It results when heat or energy waves strike an object and are converted into heat. The energy we receive from the sun is a good example. When you are wearing dark clothes on a chilly day, you may become uncomfortably hot. The sunshine warms you even though the air around you has not been heated. Radiant energy has a way of being absorbed by dark objects and reflected by light colored or shiny surfaces. Did you ever notice how snow melts faster on a black top road than it does on a concrete road?

The electric heat lamp is one of the most familiar sources of radiant heat. Other examples are panels and cables that are built into the walls and ceilings of homes to provide heat.

Make Popcorn 3 Ways

How do you make popcorn? Did you know that you can do this kind of a heating job three different ways?

Materials Needed

Popcorn
Cooking oil or shortening
Salt and butter
4-qt. saucepan, with cover. (A glass cover is preferred.)
Potholder
Electric range
2 250-watt heatlamps
2 spring clamp type lampholders
Wire mesh corn popping basket or wire mesh kitchen strainer (improvise a screen wire cover)

First, make popcorn the way you usually do. Set a front surface unit control on the range at "medium high". Pour enough oil to very lightly cover the bottom of the pan. When the pan is hot, pour in enough popcorn to cover the bottom with one layer of kernels. Use the potholder in one hand to hold the cover on, and with the other move the pan back and forth across the unit. When the popping stops, remove from the heat.

How did the heat get to the popcorn?

Second, make popcorn in the oven. Add the oil to the pan, cover it and put it in the oven. Turn the oven on, with the automatic control set at 400°. When the oven indicator light goes off, this means that the proper temperature has been reached. With the potholder, remove the pan and add one layer of popcorn kernels. Replace the pan in the oven. When the popping stops (listen for it) remove the pan.

What kind of heating took place here?

Third, make popcorn with the heat lamps. Clamp the lampholders to the back of a chair or other vertical support. They should be 6 to 8 inches apart and pointed directly at each other. Put about 2 tablespoonfuls of popcorn in the Wire basket or strainer. Do not add oil. Hold the basket midway between the two lamps. When the popping stops, turn off the lamps.

What kind of heating was this?

Now, butter and salt the popcorn you have made and share it with others.

What Did You Learn?

1. How is heat transferred from one body to another?

2. Could chicks or pigs receive warmth from a heat lamp without the air in the pens becoming warm? Explain.

3. How does a broiler unit in a range cook meat?

4. How does an oven bake food?

5. Tell why iron picture wire was used instead of copper wire for your heating demonstration.

LESSON NO. B-13

Credit Points 2

MYSTERIOUS MAGNETISM

In ancient times, people found certain rocks that clung together in bunches. These rocks were very mysterious. People didn't understand them and many superstitions grew up about lodestones, as these rocks were called. Lodestone (sometimes spelled loadstone) means leading stone. People even told Columbus not to sail out of sight of land because a giant lodestone was just over the horizon waiting to pull all the nails out of his ships.

The Chinese were the first to use magnets. They found that if you hung a lodestone by a string, one end of the stone would always point in the direction of the North Star. They had the first magnetic compasses.

An artificial magnet can be made by stroking or gently rubbing a piece of steel with a lodestone. This piece of steel then can be used to magnetize another piece of steel. This can be continued on and on. Lodestones are not always available but you can get the same results with an electric current. So, magnetism and electricity are very closely related.

What to Do

Learn about magnetism by doing the experiments that follow.

Seeing is believing!

Materials You Will Need
    2 dry cell batteries (#905)

A few feet of No. 18 bell wire
3 steel knitting needles or similar hard steel
2 ft. of light thread
Sheet of light cardboard or stiff paper
Permanent magnet (bar or horseshoe)
Compass
1 or more large nails or spikes
Red and black china-marking pencils or crayons

Iron filings
Wire cutters
Carpet tacks

(Iron filings usually can be found under the grinding wheel in a shop. If you can't find any, rub some steel wool pads together to produce bits of metal that will do.)

"See" a Magnetic Field

Cover the permanent magnet with the cardboard or paper. Sprinkle iron filings on the paper. Tap the paper and note the pattern formed. Strings or lines of filings pass from one pole of the magnet to the other. The area covered by the filings is the center of the magnetic field. To remember this, you might compare the magnetic lines of force that arrange the iron filings to the contour strips in a farmer's field.

This magnetic field is one of the important things in our everyday life with electricity. If it were not for the magnetic field, we would not have electric motors. Telephones, radios, television, and many other things we use every day also depend on this magnetic field.

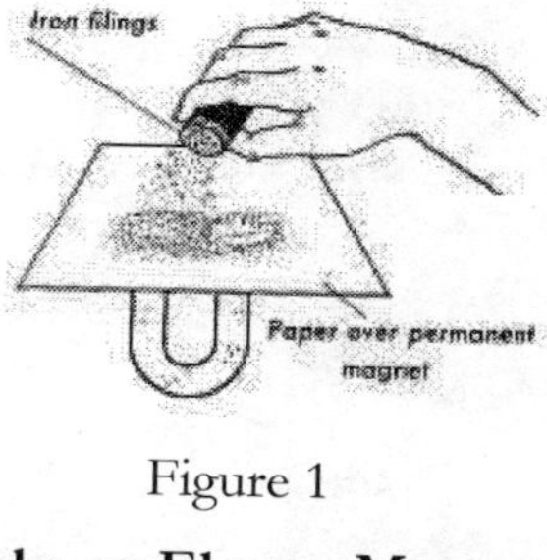

Figure 1

Make an Electro-Magnet

You can make magnetism work for you by winding several turns of insulated wire around one or more large nails or spikes (soft iron). Connect one end of the wire to the battery. Touch the other end of the wire to the other terminal for a few seconds and see how many tacks you can pick up. Repeat the experiment using as many turns as possible. How many more tacks were you able to pick up?

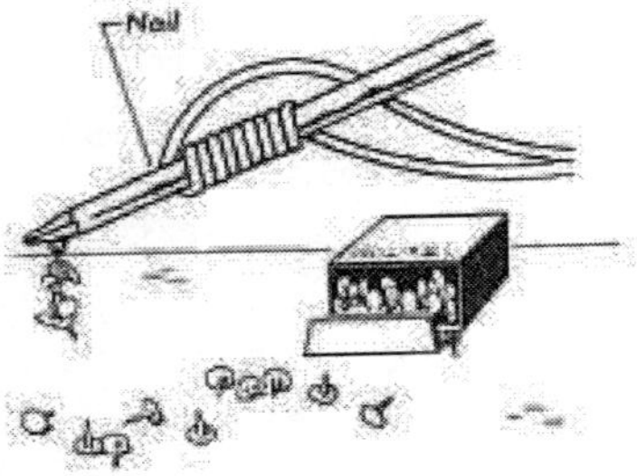

Figure 2

You have made what we call an electromagnet. When you disconnect the wire, the nails fall off. This is one of the advantages of an electromagnet. We can turn magnetism on and off as we wish. Picture a crane operator throwing the switch and picking up scrap iron and steel. Then he opens the switch to drop the scrap metals.

Soft iron can be magnetized easily as you have just seen, but loses its magnetism in a short time. Steel is harder to magnetize but holds its magnetism almost indefinitely.

Make a Permanent Magnet

Wrap the insulated bell wire around the steel knitting needle. The wire should be wrapped the full length of the needle. One end of the wire is connected to the battery. The other end of the wire is then touched for just a few seconds to the other terminal. This should make the needle into a permanent bar magnet. If you did not get results, try two batteries in series, wind more turns of wire on the needle, and leave it connected a little longer. Do the same thing with the second knitting needle. In the same way, you can magnetize a screwdriver, so that you can use it to pick up and hold steel screws. Don't do it unless you want your screwdriver to be magnetized.

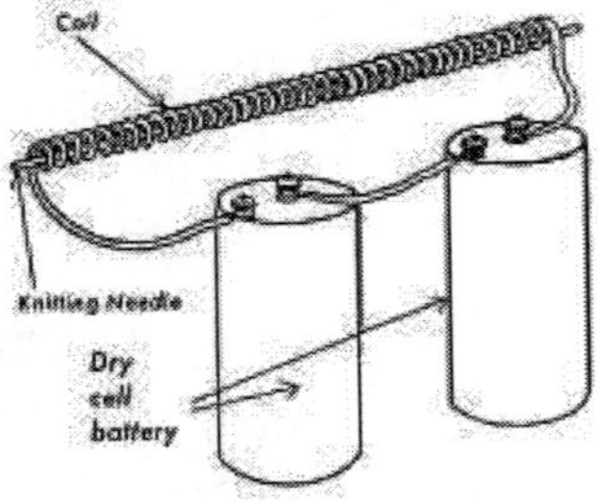

Figure 3

See How They Attract and Repel

Take one of the magnetized needles and hang it with a thread. A thread stirrup (Figure 4) will help keep it level. Be sure it is not near other large pieces of steel. Watch the needle. Does it settle down, pointing in one direction? (Check to see if this is the same direction as your compass). If it does, you have made a compass. The tip of the needle pointing north is called the North Pole (North-seeking pole). The other end is called the South Pole. Mark the North Pole with a stroke of the red marking pencil. Mark the South Pole black. Do the same thing with the second needle. You can show this with a sewing needle, and a notched cork, and a bowl of water. Rest the needle in the notched cork, and float it on the water.

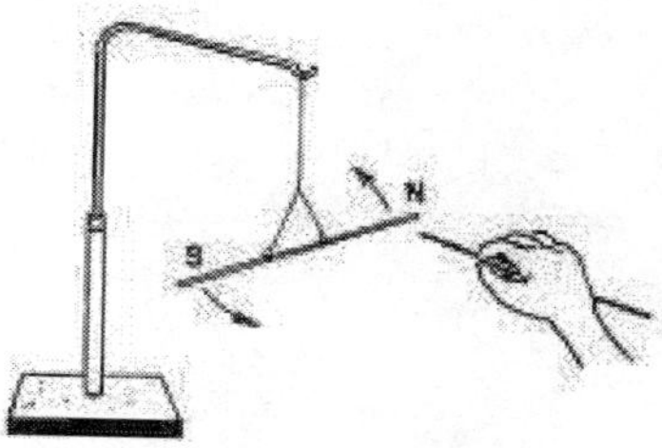

Figure 4

Hold the compass near the North Pole of the needle. What happens? Does the South Pole of the needle attract the North or South Pole of the compass? Try this with the second magnetized needle. See if you can prove the rule that like poles repel (drive away) and unlike poles attract.

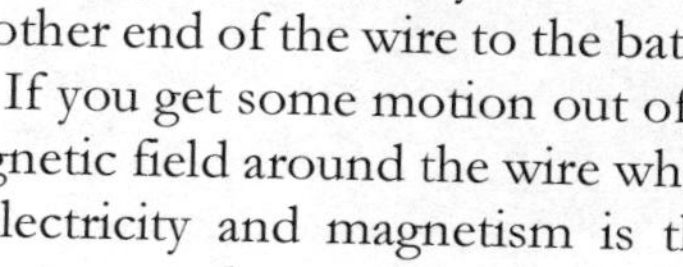

Figure 5

Connect one end of a wire loop to the battery and run the wire directly over the compass. Touch the other end of the wire to the battery. Which way does the compass point now? If you get some motion out of the compass needle, this proves there is a magnetic field around the wire when current is flowing. This relation between electricity and magnetism is the thing that makes electric motors and generators work.

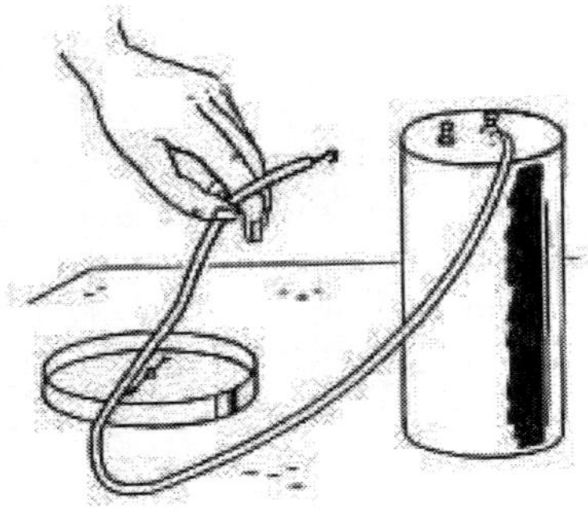

Figure 6

Make Many From One

Lay the third needle (unmagnetized) on a table and stroke it with one of the magnetized needles. (See diagram) Always stroke it in the same direction. Raise the magnetized needle at least two inches on each return stroke. Thus you can magnetize the needle by using the other needle.

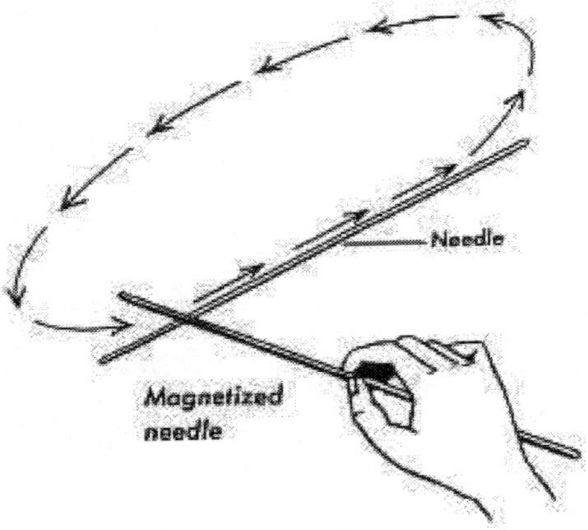

Figure 7

Use the wire cutters to cut the first magnetized needle in short lengths. (Cover the needle with a cloth to keep the pieces from flying.) Can you show by using the compass that each piece is a complete magnet? Hold one end, then the other, of each piece to a compass. Does each piece have both a North Pole and a South Pole?

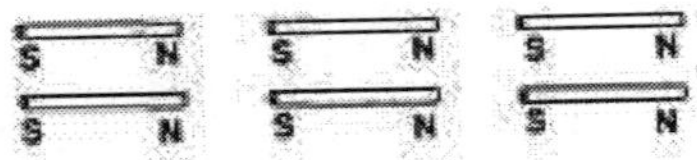

Magnetism and Animals

The things you have done show that electricity and magnetism are related in many ways. Magnetism is mysterious, and there are still things to discover about it. It is thought that animals and birds are aided in their sense of direction by magnetism. It is commonly known that when a person gets lost in the woods, he tends to go around in circles. Possibly this is caused by the earth's magnetic field.

What Did You Learn?

1. Where are natural magnets obtained?
2. How can artificial magnets be made?
3. What material is needed for a permanent magnet? For a temporary magnet?
4. How can you find out which is the North Pole of an unmarked magnet?
5. How many poles does a magnet have?
6. Which magnetic poles attract each other?
7. Why couldn't you make a compass out of a strip of plastic?
8. What causes the compass to change direction when a wire carrying battery current is held over the needle?
9. List the materials you would need and tell how you would build a homemade compass.
10. Tell what you enjoyed most about becoming acquainted with mysterious magnetism.

LESSON NO. B-14

Credit Points 2

Give your appliances and lights a square meal

Would you say that having enough to eat was pretty important in the home that you know?

The "food" for your appliances and lights is electricity, and like you they must be "fed" enough.

What to Do

1. List the appliances and lights in your home.

2. See if any of them are "starving" for the electricity they need.

3. Learn how the electricity gets to where it's used.

4. Make a chart of the electrical circuits in your home.

5. Make sure that each circuit is protected with the right fuse or circuit breaker.

Count Your Electrical Blessings

Many people in much of the rest of the world wish that they could trade places with us, because we have so many electrical appliances in our homes.

Of course, we have not always had as many appliances as there are today. When electricity first came along, people used it only for lights. Then, they began to add flatirons, washing machines, refrigerators, coffee percolators, and radios.

Then more and more electrical things were made for people to use and enjoy. Now we have dozens and dozens of uses for electricity in our homes.

How many different uses for electricity are there in your home today? Ask your parents how many there were when your home was built or first wired. How many were *common* when your parents began to keep house?

Some Homes Are Behind Times

Many older homes were built before electricity was available, and were wired later. And like them, some older homes that were wired as they were built had only enough wiring for lights and a few other appliances, because those were the only uses that were known at that time.

But people kept on living in these homes, and kept adding to the uses they made of electricity without adding to their wiring.

What has this meant? Well, if electricity were like cars and trucks, you could say that some people are trying to put turnpike traffic through a back-country dirt road!

Watch for Signs of Starvation

Of course, as your state has done with its highways, some people have expanded and modernized their wiring. But many others have not yet seen this need, or if they have, they may have to do it again.

Here's why:

Your power supplier delivers current to you at the right voltage or electrical pressure. If the wires in your house are large enough, they will pass this full voltage on to the appliances.

But if your wiring is too small, the electricity arrives at the appliances so weak that they can't work properly, and much of what you pay for is wasted.

Here are some things you can watch for in your own home. They will tell you whether your appliances are getting enough electrical "food" or not.

1. *A shrinking TV picture*—If it draws in from the sides of the screen, fades, loses contrast, or if the sound becomes distorted, you may have low voltage.

2. *Too much fuse blowing or circuit breaker tripping.*

3. *Heating appliances are slow to do their jobs.*

4. *Lights dimming*, when motors or other appliances are turned on.

There Should Be Enough Ways to Get "Appliance-Food" Around

If appliances in your home show these starvation signs, then you may not have enough ways for the electricity to get to where it's used.

There are three kinds of these electrical highways or circuits, and your home should have enough of each:

1. *General purpose circuits*—These serve lights all over the house, and convenience outlets everywhere except in the kitchen, laundry, and dining areas.

A rule-of-thumb is: There should be at least one general purpose circuit for each 500 sq. ft. of floor space.

2. *Small appliance circuits*—These are not used for lights, but instead they supply convenience outlets in the kitchen, laundry, and dining areas where portable appliances are most used.

Every home should have at least two small-appliance circuits.

3. *Individual or special-purpose circuits*—One of these is needed for each: electric range, dishwasher, water heater, freezer, automatic washer, clothes dryer, air conditioner, pump, and house heating equipment.

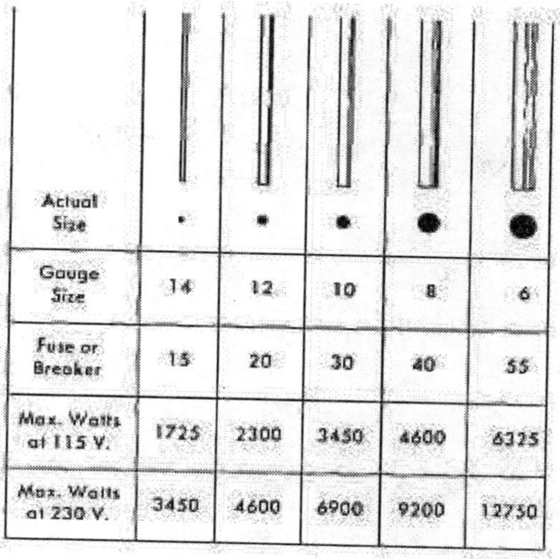

Actual Size					
Gauge Size	14	12	10	8	6
Fuse or Breaker	15	20	30	40	55
Max. Watts at 115 V.	1725	2300	3450	4600	6325
Max. Watts at 230 V.	3450	4600	6900	9200	12750

Wire sizes commonly used in homes

Each Circuit Big Enough

The capacity of each circuit is limited by the size of its wires. The chart above shows you the actual sizes of wires commonly used in permanent home wiring, and what each will carry. Notice that each size is given a number, and the smaller the number, the bigger the wire.

Also notice that a given size of wire will carry twice as many watts at 230 volts as it will at 115 volts. (Watts are figured by multiplying amps times volts.)

General purpose circuits usually are either Number 14 or Number 12 wire, at 115 volts. What is the capacity of each, in watts? (Number 12 wire is recommended for all new general purpose circuits.)

Small appliance circuits are required to be at least Number 12 wire.

Individual circuits are always sized according to the appliance they serve. Find the size wire that should be used for a 10, 000-watt, 230-volt range; a 1500-watt, 115-volt dishwasher; a 4500-watt, 230-volt clothes dryer. ________ ________ ________

Only One Fuse Size Right

A fuse in an electrical circuit is like an alert traffic policeman—stopping everything if there's danger. A circuit breaker serves the same purpose, and the right size is installed when the wiring is done.

A policeman uses his brain to tell him when to blow his whistle, but a fuse depends on the size of the little fusible (meltable) metal link that you see under the glass.

If too great an electrical load is added to a circuit, this link will melt and prevent a dangerous overload. If you put in a fuse with too heavy a link, it will not melt in time, and the wiring and equipment may be damaged.

Therefore the right size of fuse is very important, and is something that you should check in your own home.

See the chart above for the right fuse for each size wire.

Make a Circuit Chart

At one or more places in your home there is a box or panel containing the fuses or breakers for the various circuits. Attached to the inside of the door of each such panel should be a chart something like this:

No	Description	Fuse size
1	Range	40
2	Kitchen Outlets	20
3	Dining Room Outlets	20
4	Living Room Outlets	15

Notice that in our chart we have made columns for a description of what each circuit serves, its number or position in the panel, and the proper size fuse for it.

Because most such charts leave out this last very important bit of information, you should make a complete new chart, like the one shown. Provide as many lines as there are fuse positions. Paste or tape it to the inside of the panel door.

Then, ask permission of your parents to disconnect all the circuits by unscrewing the fuses or flipping the circuit breakers. *Do not touch anything but the fuse rim.* Then reconnect them, one at a time, to find out what each circuit serves. Turn on as many lights as you can, to help you in your detective work. Use a test lamp at those outlets that do not have a light connected to them.

Write two or three words describing each circuit on the proper line on your chart.

On a separate sheet, keep track of the appliances and lights that are on each circuit, and add up the watts. (If the name-plate of any appliance gives "amperes", "amps", or "A" instead of watts, just remember that amps times volts equals watts.) This will tell you if any of them are overloaded. Show this sheet to your parents.

Check the Wire Sizes

Disconnect the main switch, and determine the size of the wires in each circuit. Don't include the insulation in your measurement.

BE CAREFUL!

Even though you have disconnected the main switch, the wires coming into it are still "live". So, do not touch any wires. Instead hold the wire size chart near them so that you can tell which gauge each one is.

Write in the proper size fuse for each circuit on your chart.

Replace Any Wrong-Size Fuses

Do the fuse sizes you have written on your chart agree with the ones that are in place in the panel?

Get the right size fuses and replace any that are wrong. Make sure that you have a reserve supply of the right sizes, and that they are handy for future use.

Talk it Over With Your Parents

Do you think that your home has enough of the proper size circuits? If not, talk it over with your parents. They may want to ask an electrician to go over the wiring and make the necessary changes.

What Did You Learn?

(Underline the right answer.)

1. A (television set, radio) is very sensitive to changes in voltage.

2. Dimming lights mean (static in the wires, an electrical overload).

3. Wires that become warm from overload make it (more expensive, cheaper) to operate the equipment.

4. A home of 2,000 sq. ft. should have at least (three, four) general purpose circuits.

5. One solution to low voltage symptoms is (heavier fuses, more circuits).

6. Full capacity for a Number 14 wire circuit at 115 volts is (1725 watts, 3000 watts).

7. A room air conditioner should be on (a general purpose, an individual) circuit.

8. The purpose of a fuse is to (let you disconnect the circuit, automatically prevent overloading the circuit).

9. The right size fuse is determined by (wire size, the store where you buy it).

10. A circuit chart should give (circuit description and fuse size, the maker's name).

Demonstrations You Can Give

Ask your leader to help you plan a demonstration. You can show how lights dim when too many other appliances are connected, how a fuse protects against overloading, and the danger of using too large a fuse.

For More Information

Ask your Extension agent, power supplier, or electrician for additional help.

LESSON NO. B-15

Credit Points 4

YOU CAN MEASURE ELECTRICITY

Instruments that can detect or measure the flow of electricity have helped to make possible the wonders of electricity as we know them today.

Scientists in laboratories must have measuring devices for experiments leading to new uses of electricity. Power suppliers must have instruments that tell what the generating equipment is doing and to measure the amount of electricity being sold to users. Factories need instruments that keep tab on electrical equipment to make sure electricity is being used efficiently.

In fact, almost anywhere you find electric power at work you'll find electrical instruments—even in your home. The one you know best measures the amount of electricity used. Another, in the family car, shows whether the generator is charging the battery or if the battery is discharging.

What to Do

1. Make a simple kind of direct-current meter that will show you that there's a magnetic field around a wire carrying an electric current and that will detect a very tiny current.

2. Make a more refined D.C. instrument (galvanoscope) and measure the voltage of different sizes of dry batteries, and show how an electric current can be induced.

Tools and Materials You'll Need:

Pair of pliers, knife, small hammer
30 feet of No. 24 bell or magnet wire
Compass
Two coins—a penny and a dime
Fine sandpaper

Blotting paper
Plastic or cellophane tape
Wooden blocks (See Figure 4)
Glue
2 small nails
One #905 dry cell, a penlight battery, and two regular flashlight batteries
Table salt
Drinking glass
2 paper clips
Two machine bolts

How They Work

Like many electrical things, most electrical instruments depend on the action of magnetism created by an electric current. There is a magnetic *field* or lines of force around any wire carrying an electric current. If this field is controlled and made to react on a sensitive device, like an easily moved pointer, we have an electrical instrument.

Detect a Magnetic Field

First, let's prove that there is a magnetic field around any wire carrying an electric current. Take a piece of wire about two feet long and scrape off about an inch of insulation from each end. Connect one end to a battery terminal. Make a loop of wire that crosses the face of your compass, north to south. Now touch the other end of the wire to the other battery terminal.

(DO NOT attempt to substitute alternating current, as from a model railroad transformer because its alternating current will cause the compass needle to swing rapidly from one side to the other.)

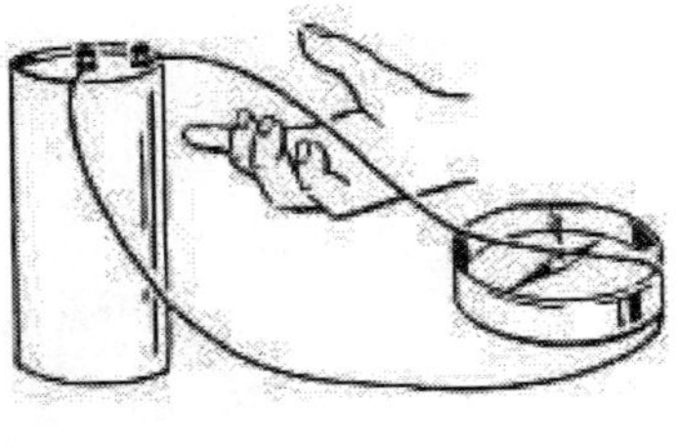

Figure 1.

Put your right hand beneath the wire so that your fingers point the way the needle deflects, and your thumb will point in the direction that the current is flowing.]

What happens? Your compass needle should move to one side because it is very sensitive to magnetic influences. This proved that the wire created a magnetic field or lines of force when we passed electricity through it. (Figure 1)

Detect a Tiny Current

How sensitive is your simple electric meter? Take about five feet of wire and wrap it around your compass as in Figure 2, keeping the turns bunched together as much as you can. Leave about six inches at both ends of the wire extended for leads. Scrape the insulation off the last inch of both. Rotate the coil and compass until the needle and coil are parallel, both pointing north and south.

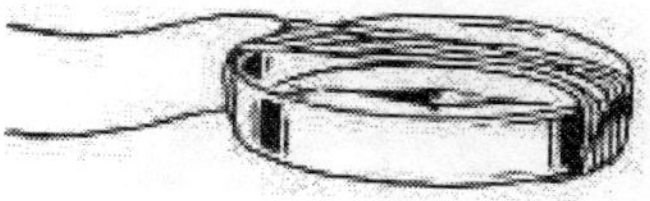

Figure 2

Take a copper penny and a dime, and clean off any corrosion or film on the coin faces with a bit of fine sandpaper. Now take a piece of blotting paper about the size of the penny and dip it into strong salt water. Place the damp blotting paper between the penny and the dime. Place one of your compass coil leads against the dime, and the other against the penny as shown in Figure 3. Be sure you have good metal-to-metal contact between the wires and the coins.

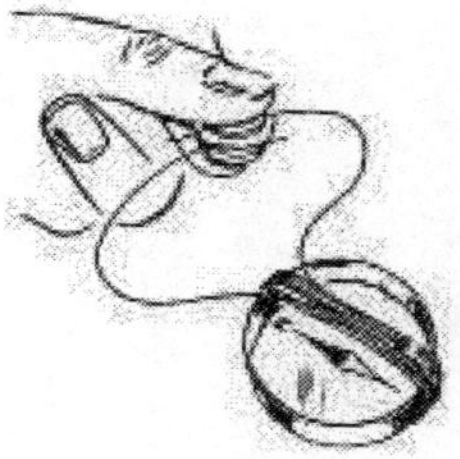

Figure 3

At the instant that you squeeze the leads against the coins, watch what is happening to the compass needle. It should move for an instant from the north position each time you press the leads against the two coins.

Obviously, the little coin battery you have just made produces a very weak electrical current. Even so, your instrument should be able to detect it.

Make a Simple Galvanoscope

Now let's make a meter that is a little more practical to use. Broadly speaking, a galvanoscope is an instrument that detects the presence of electric currents.

It sounds complicated but it is really quite simple. It is named in honor of an Italian professor named Galvani who made important early experiments with electricity.

A refinement of the galvanoscope is today's galvanometer. Other related instruments are the voltmeter and ammeter. These are very important instruments to the electrical engineer.

Using a glass or anything three to four inches in diameter, wind about 20 turns of wire in a "bunched" coil as in Figure 4. Wrap the coil at several points with cellophane or plastic tape to keep it from unwinding.

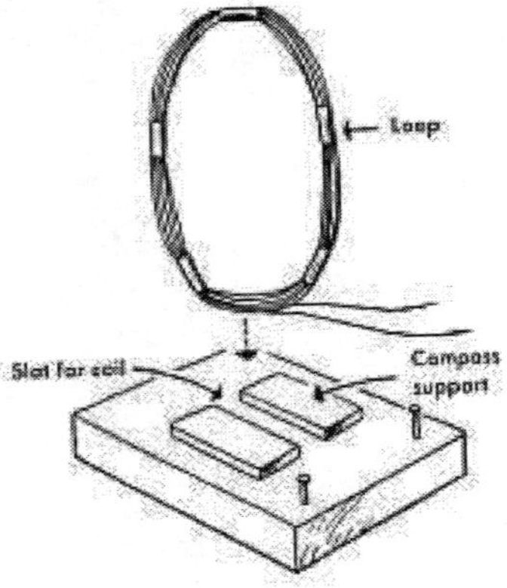

Figure 4

Make a wood base for your coil as shown in Figure 4. The compass support blocks can be thin wood slats. Do not attach them with steel nails or tacks. Use glue instead. Hold the coil in the slot between the blocks with glue or melted wax or use copper staples. Place the compass on the supports and rotate the base so that the compass needle and coil are parallel, pointing north and south.

Measure the Voltage of Batteries

Do you know what difference the size of dry cell battery makes in the voltage it supplies? Your meter can tell you.

To test the voltage of batteries we must be able to control our galvanoscope. To do this, connect a glass of strong salt water in series with the battery as shown in Figure 5. Make sure the wire ends immersed in the salt water are scraped free of enamel.

Figure 5

With one of the batteries connected, move the wires in the salt water first closer, then farther apart (keeping them parallel to each other) while watching your compass needle. When the needle stays 15 to 20 degrees off north, lock the wires in the salt solution in place with paper clips.

Now disconnect the battery you have been using and connect a smaller battery. If both batteries are fresh, the compass needle should return to almost the same spot. This proves that both batteries regardless of size put out the very same voltage. The larger ones, however, are designed to last longer.

Measure the Difference between Series and Parallel

Using the salt solution as in the previous experiment, connect two flashlight batteries in series as shown in Figure 6. The compass needle should move about twice as far as it did with one battery connected. This shows that when you connect batteries this way you double their voltage.

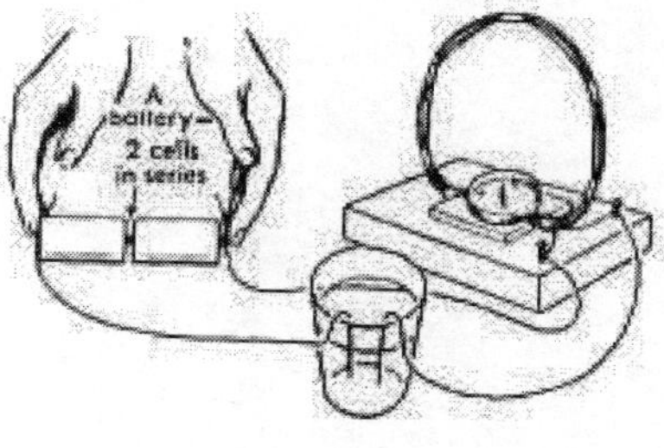

Figure 6

Now place your batteries side by side and connect the two top terminals and the two bases as shown in Figure 7. The compass needle should move only as much as it did for one battery. This is called a parallel connection. You can see that this arrangement does not double the voltage, even though you used two batteries.

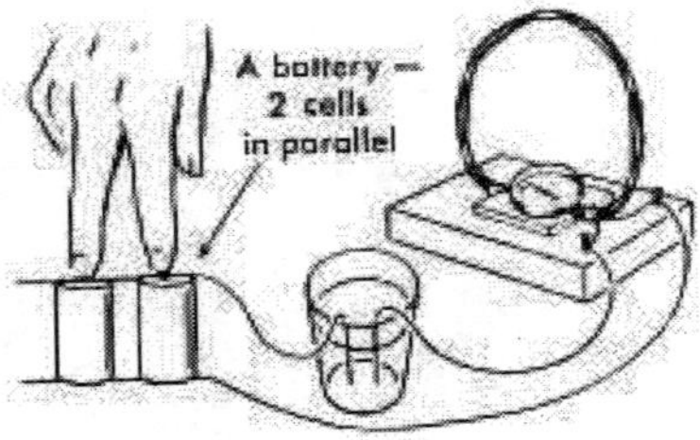

Figure 7

While you have this hookup, try reversing the position of the leads connected to your batteries. Notice that reversing the direction of current flow in the coil causes the compass needle to swing in the opposite direction.

Test for Induced Current

Make a simple coil by winding about 50 turns of wire around a machine bolt core. The bolt should be 1/4 to 1/2" in diameter and about two inches long. Connect the coil to your galvanoscope as shown in Figure 8. Pass the coil back and forth close to the end of a permanent magnet.

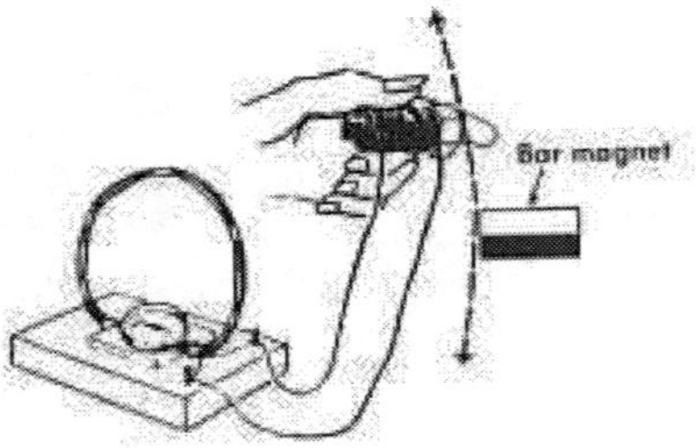

Figure 8

Notice a slight deflection of the compass needle with each pass. You have shown that electricity can be induced in a wire coil by moving it through a magnetic field. Currents generated in this way are called induced currents.

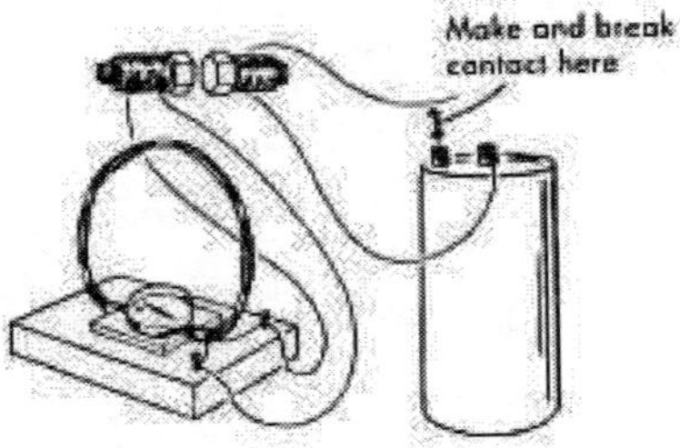

Figure 9

Now make another coil and core just like the first one and arrange them and a connection as shown in Figure 9. If you make and break the current to the second coil, you will build up and collapse a magnetic field around the first coil and again induce a current in it. You will see the compass needle swing back and forth again.

These last two experiments give you a crude idea of how an electric generator works, producing electric current by induction as a coil-wound rotor revolves within a magnetic field.

What Did You Learn?

What does every current-carrying wire have around it? How does this help us to measure electricity? How sensitive are electrical instruments? What is the difference in voltage between (a) a large and a small dry cell? (b) batteries connected in series and in parallel? (c) your original connection and the reverse of it? What similarity does the test for induced current show between movement through a magnetic field and the making and breaking of a direct current?

Demonstrations You Can Give

Show others how your galvanoscope can detect: whether a battery is producing current, which way the current is flowing, and whether a current is strong or weak. Demonstrate how a current can be generated using magnetism.

For More Information

Ask your power supplier representative to show you some of the instruments used by his organization, and to give you a brief explanation of how they work. Ask him or an electrician to give you a demonstration of a split-core ammeter.

Made in the USA
Monee, IL
05 November 2026

67380108R00062

Epopee riscoperte
vol. VI

Il risveglio di Fufluns

Il vino etrusco contemporaneo
e la chimera delle origini della viticultura

Appunti di ricerca esistenziale

Edizioni Press & Archeos
Collana *Epopee riscoperte, VI*
via Cittadella, 9
50144 Firenze
http://www.pressandarcheos.com
info@e-archeos.com

prima edizione, primo aggiornamento, dicembre 2017

Grafica di copertina e ricerche iconografiche a cura dell'autore

Questo libro è dedicato a

Marco

bevitore, ricercatore, amico

Prefazione - enosofica

...La moderata caparbietà dei profumi di frutti di bosco dei sangioveti che accendono l'estro di Tinia e il desiderio di declinazioni fondamentali; il rosso rubino e i sentori di viole, more, marasche dei vini di Tuscia che destano la tufacea durezza di Tages e le difese dei tesori sotterranei; l'avvolgente essenzialità della ginestra e della mammola dei Chianti che cerchiando i rimossi dei dolcicolli attraverso ironici torrenti e filari, provoca i sorrisi di Turan e le sottili trame di Uni; i sentori selvatici misteriosamente trasmessi dalle vicine macchie ai nettari delle ventose maremme, l'umore del sempervirens, *la gemma di lucertola, l'alito di cinghiale che attira le nude schiere di Artume; la sfida aromatica pronunciata, nella danza con i vitigni autoctoni - tra lamponi e gelsi, quercia e pepi, Appennini e Borgogne - dai Pinot d'ogni sorta che chiedono a Voltumna la mediazione con il* pantheon *del Nord; l'inchino bordolese preteso dai lucumonici Cabernet, con i loro mantelli di Verdot e paludamenti di Merlot, gli smeraldini registri di sentori, le chiocce dorate nascoste in tombe inviolate...e così un'infinità di visioni, suggestioni, fantasticherie ma anche strutturalità e funzionalità; ecco, tutto questo o molto altro il* vignaiuolo, *nel profondo del gioco della sua missione, vorrebbe farlo proprio, domarlo e donarlo attraverso un naso ideale ereditato da qualche nume, applicando una tradizione universale quanto frugale: rileggendo manoscritti di fattori settecenteschi lucidamente insòbri, immaginando formule propiziatorie aruspicine, ascoltando racconti d'anziani in odor di terra, fraseggi che parlano di lune variabili, di santi bagnati o arrabbiati, di fulmini ad Est e di silvestri prodigi; d'alberi fruscianti intermittenti come oracoli, di capsule villanoviane emerse nei sodi maledetti, di draghi sotterranei che russano fuoco*

sotto vigneti di malvasia, di orsi stellati tra castelli d'alveari...il cui miele ricompare negli stessi vinsanti; vorrebbe raggiungere e realizzare un segreto enosofico che è spirituale e al contempo sensoriale, essenziale nel velluto superficiale, uno e altro, uinu *e* vinus, *trascendente nelle acque mercuriali stillate da galestri parnasici eppur familiare, basico, dolce come un saggio che scherza con l'ultimo dei suoi seguaci.*

E dunque, insieme a costui sogna anche il bevitore, amico, parente o adepto che scivolando-nelle-ascese dell'ebbrezza ricarica il proprio io, lo rende luccicante, inscena un intuito danzante, brinda imbarazzato a paesaggi enigmatici, simula - tra sé e sé - moderate sindromi di Stendhal innanzi alle etichette, sperimenta nell'effigie ciò a cui la fortuna non ha ancora offerto appiglio, gioca con gli arabeschi della propria anima e con i nessi della propria ombra: riscopre il futuro, incalcolabile, che accogliente gli sorride...come in un antico Oriente.

Ed insomma lui e l'altro, loro e noi, accomunati forse nella figura di un cosmico attonito spettautore, sappiamo che quel contenitore perfetto, scrigno aromatico, duplicato di profumi può conferirci dignità e un qualche prestigio; sappiamo che quella recondita chimericità, quella ruspante commozione, quella disposizione d'ebbrezza hanno una forza che può eccitare, circuendo gli animi, attraverso le visioni ridipinte, nelle etichette, spargendo occasioni ad artisti vagabondi; e possiamo esser noi, vignaiuoli e bevitori, gli artisti e condire queste visioni con le nostre immagini: dei volti e nomi dei nostri amici e amori, dei desideri e degli abissi, della luce seducente dei precedenti fallimenti se non di quella del mistero di un diacronico successo...di cui s'accarezza il sogno come bambini che fantasticano osservando riflessi di luce in un prezioso,[1] *dando un nome a questo progetto, a questo eroe, a questo amico invisibile: il* daimon *di quello*

---

[1] Difatti, per questa introduzione, è a Claudel che mi sto ispirando. Paul Claudel, *Mistica delle pietre preziose*, Sellerio, Palermo 1991.

stesso Vino.

E questo nome, nel caso della nostra ricerca, è generalmente il nome di un etrusco o ha a che fare con un concetto etrusco.

Cosa comporta tale evidenza? Cosa fa l'Etrusco agli uomini di cui è "socio" e amante, nesso psichico e appiglio commerciale? e cosa fa al suo Vino stesso, alla sua sostanza divina, alla sua essenza o, più cinicamente, al suo marketing? *In questo libro, un po' diario e un po' saggio, drammaticamente lucido e pure un po' ebbro, proveremo almeno ad andare più a fondo…rischiando qualcosa di più intimo ma non per questo solo personale.*

«...Il campo ti darà viti fluenti e fornitrici di molto vino, questo ti sarà fertile d'uva e di liquido che beviamo dalle coppe dorate, non appena il grasso etrusco addetto alle are soffia nel flauto d'avorio e offriamo su larghi piatti le viscere fumanti...»

Virgilio
(*Georgiche,* II,193)

Il senso esistenziale di un *bere* etruscofilo

In questi giorni pescati da anni confusi, nel corso di una crisi che ha investito e rimescolato ogni livello e settore creativo/produttivo, con l'essenza della *terra* che continua a nascondersi e l'*uomo* che prosegue nel suo camuffamento, ho deciso d'interrogarmi su una delle esperienze più autentiche che può capitare ancora di fare: il *bere*. È una scelta per certi versi emozionale, richiede forse una presa di distanza dalla rigorosità della ricerca storica aprendo ad una complessa poetica del reale; ma le relazioni con il territorio e con la società fanno del *discorso sul vino* una sonda speciale, articolata e magnetica, utile a raccogliere un fascio d'importanti tendenze e tensioni.

I vini di cui propongo al lettore d'occuparci sono quelli che si orientano esteticamente - per vocazioni similari ma attraverso scelte alquanto eterogenee - sull'immaginario dell'antico popolo degli Etruschi.

Grazie a questo nostro speciale oggetto d'indagine non sarà difficile, nel corso della ricerca, saltare da argomentazioni che riguardano l'antichità ad altre attinenti a realtà produttive contemporanee: sono in un certo senso gli stessi *vini etruscofili* a richiederlo. D'altronde, favorirli in questa loro tendenza (e a volte "presunzione") è un po' come metterli alla prova, misurare la consapevolezza della loro stessa genesi.

Sarà necessario guardare in modo attento all'antichità ma, dopo secoli d'evoluzioni scientifiche e stratificazioni culturali, non potremo dimenticare quanto sia cambiata la sostanza e l'immagine del *vino;* e come questo cambiamento resti per un verso insondabile, poiché ci mancano gli appigli per una piena comprensione del rapporto tra *bere* e individuo nel passato più remoto. Se dunque produzioni vinifere, convenzioni, cerimonie e riti si negano in gran parte alla luce della nostra razionalità, possiamo intuire che quel *patto etrusco* a cui "chiedono" d'esser riferiti certi vini contemporanei corrisponda, in buona parte, ad un'entità di sintesi attuale, psicologicamente viva che solo si ispira, nelle sue forme (cioè nelle immagini e nei nomi) all'antico popolo e che in realtà ci parla anzitutto della nostra collettività, attuale, nazionale ma soprattutto regionale e sub-regionale; dei suoi valori e dei suoi arcani, delle sue risolutezze e delle sue problematicità.

D'altra parte questo fluttuante fondamento, cioè il rifarsi agli Etruschi e ad un atavico senso di collettività, non è mai completamente chiaro e raramente è letteralizzato. Degli Etruschi chiamati in causa pare secondario conoscere con precisione la realtà storica, le autentiche motivazioni che sembra restino confuse tra soluzioni fantasticose, paganeggianti e a volte sepolcrali, fondendosi con l'inconscio - e quindi con gli approdi e le derive personali - del bevitore quanto, in certi casi, del *vignaiuolo* stesso. Ma forse anche per questo il *marketing* del vino adotta volentieri le "forme etrusche" che in modo così spontaneo si radicano in profondità.

Sembra altresì plausibile come il rimando agli Etruschi possa celare, appunto, la realtà psicologica di un *patto* magico/misterico, riacceso in seno alla complessità storica e iconologica delle "etruscherie" - come venivano chiamate nei secoli addietro, quando le prime sepolture principesche, con i loro corredi, producevano lo stupore (credo mai estinto) per l'esistenza di un grande popolo pressoché rimosso dalla storia -. Talché sembra impossibile svelare il fascino dei *vini etruscofili* senza confrontarsi con decisione con il fascino degli Etruschi e del loro vincolo sacrale.

La viticultura etrusca, riproiettata nel presente, tra le sue possibilità concettuali e mai pienamente imbracciate sembra rilanciare, per così dire, l'idea di una possibile *nazione* del vino quindi l'essenza di un popolo che ancor oggi può riconoscere sé stesso - e una sua pur vaga identità - nella civiltà della produzione, del consumo, del commercio, della sacralizzazione del vino. A pensarci il "fattore vinifero" unisce tre o quattro regioni italiane più di quanto la politica e la cronologia storica siano riuscite a fare. Pur essendo pressoché utopica, l'idea di un *popolo del vino* di Tuscia è avvertibile-intuibile in Chianti come nel Viterbese, nei Colli aretini come in Val di Cornia, nell'immediato nord di Roma come nell'Appennino tosco-emiliano...

Il membro di questa *collettività* la cui vita parrebbe svolgersi tra ovattati banchetti, nel presunto "agio etrusco" (una percezione-pregiudizio con cui conviviamo da più di duemila anni) e in un territorio che sembra facilitarlo nel duro confronto con la natura, non mostra però l'intenzione di "rivendicare" il territorio ed affermare le caratteristiche etnologiche-identitarie che lo contraddistinguono: il vino stesso sembra dare soddisfazione alla pienezza del *patto* e inutile sarebbe estendere la questione a nuove e vecchie idee di legazioni, federazioni, circoscrizioni. Quella del *vino* e degli *Etruschi* sembra essere una *collettività* in un senso esistenziale, legato intimamente ai *possibili* che l'ebbrezza rende disponibili e che la produzione vinifera rende dispositivi e operativi.

Tutto questo produce forse un complesso dualismo tra la regione reale/archeologica e quella ideale/possibile. In altri contesti storici, questo genere di contraddizione è divenuto materiale per i romantici; ma non sempre ci si ferma alla filosofia: le distorsioni possono creare problemi politici, specie là dove il coinvolgimento scaturito dall'araldo della viticultura è fattore primario nel riconoscimento del popolo in sé stesso e nella sua economia. Questo è avvenuto ad esempio in Chianti dove un certo *dissidio concettuale* favorisce incongruenze geografiche ed un uso "anarchico" del nome originario («*chianti*»). Si è parlato difatti di un *Quarto Chianti* e di un vino capace di rinsaldare il legame con uno speciale "aldilà" comprensivo dei regimi immaginari delle esperienze locali, esulando da qualsiasi confine fisico - compreso quello dell'omonimo Consorzio -.[2]

Tornando al nostro soggetto, nel corso della ricerca vedremo come alcune attività e mentalizzazioni, presumibilmente "etrusche" ed oggi soprattutto immaginarie, sembrino ancora strutturate per svolgersi in un territorio sintetico, o proprio *mitico*, che il vino stesso può rivelare e rendere, in qualche modo, praticabile. Ed è altresì possibile intuire come, per gli antichi, la differenza tra questa dimensione ideale e la realtà quotidiana fosse assai più limitata di oggi: quindi il vino poteva realmente far da ponte-*medium*, diciamo, per il divino - comunque per un livello di più profonda partecipazione vitale -, poggiando con naturalezza sulla struttura del *sacro* e sui relativi attributi.

I *vini etruschi* permettono forse, grazie ad una serie di caratteristiche che proveremo ad analizzare, l'abolizione dell'impressione di una distanza da ciò che è più *profondamente riunificante*. E lasciar che si perfezioni tale abolizione è tra gli obiettivi della mia ricerca e al contempo conseguenza di un certo qual modo di ricercare - che comprende in parte, inevitabilmente, il bere.

Sia chiaro che chi scrive non è né un *sommelier* né tanto meno un vignaiuolo, ma senz'altro un bevitore attento, che ha dedicato molto tempo della sua esistenza al rapporto con le *immagini* e che quindi si muove istintivamente tra le strutture dell'*immaginario*. Premesso questo, occupandomi per lavoro anche di *marketing*, sento di poter accompagnare il lettore per qualche tratto di un *cammino*.

---

[2] Si veda anche L. Pecchioni, *Il Quarto Chianti*, Press & Archeos, Firenze 2012.

Procedendo dunque nell'osservazione di possibili tracce dei *numena* etruschi, cioè di *Fufluns* (dio del vino) e delle principali divinità nei vini contemporanei – e si tratta badiamo bene di un'osservazione partecipante, che implica una fusione e una contaminazione, quindi lucida ma non del tutto esatta, statistica o razionalistica, forse perché deve adeguarsi, a tratti, alla sfuggevolezza del suo antico paradigma – abbiamo, sulla base di quanto premesso, più sentieri da percorrere che s'aprono a proposito di un vino, di un *terroir*, di un *immaginaire* o altro.

Su tutto, risulterà di primario interesse la questione della figura o *persona etrusca* legata a specifici vini che ne carpiscono il nome, che sono *titolati* ad essa. Quanto e come questa fluttuante *persona-immagine* è oggi agente sensibile, avvertibile, insomma *reale*? E ancor prima, cosa comporta la nominazione di molti vini come "*Etrusco*", "*Rasna*", *Rasenna*" o simili? Cominceremo proprio da qui: dai vini che portano *nomi di popolo*.

L'interpretazione del potere del nome verrà poi ripresa alla fine di questo testo, una volta assunti elementi utili su tale prospettiva.

I

i Vini

Aggettivi e Nomi di Popolo

Ripartiamo proprio dal già nominato *Chianti*, per molti il cuore dell'immaginario vinifero toscano, dove troviamo alcuni casi che rievocano il *fattore etrusco* o che si definiscono letteralmente «etruschi»; segno che nel territorio è riconoscibile e riconosciuta la componente arcaica nonostante, almeno in senso turistico, si tenda ad associare il Chianti soprattutto al Medioevo.

Un caso eccellente è *Etrusco* di Cennatoio Inter Vineas, un IGT, Sangiovese in purezza dal rosso rubino intenso, equilibrato e dai profumi complessi in piena tradizione toscana; uno di quei vini che già a stappare la bottiglia, ai primi odori che si propagano nell'aria, ti senti fiondato tra le vigne e le macchie di un *patchwork* chiantigiano. L'azienda si trova sugli antichi itinerari (già frequentati dagli Etruschi) che collegano Panzano a Volpaia, zona definita un tempo Stinche: sequenza di colli dove trovarono rifugio signori - ma anche banditi e fuorusciti - d'ogni epoca.

L'Etrusco ridisegna questa complessità con scaltra personalità, stabilendo un vivido equilibrio tra acidità, tannini e alcoli. Si presenta all'olfatto con note boschive, di frutta rossa e secca «per poi aprirsi in una gamma di profumi più complessi che ricordano il cuoio, la vaniglia, il cacao e la liquerizia». [*S.*] Il finale è vellutato, con echi di spezie. Nell'etichetta, raffigurante un antico guerriero su sfondo nero, il nome del vino è scritto in caratteri etrusco/fenici, da destra a sinistra, la direzione di scrittura originaria; in realtà gli Etruschi non scrissero mai quella parola e probabilmente non si definirono mai con quel nome. Una contraddizione storica alla quale ci abitueremo.

Sempre in Chianti, un altro vino di nome *Etrusco* è prodotto

I vigneti dei pressi di San Donato in Poggio (FI) da cui proviene un «Etrusco».

a Montecchio, nei pressi di San Donato in Poggio. Confezione maestosa per un nettare dall'incedere suadente, vellutato, con finale «sapido ma senza prepotenza» [*S.*]; il Cabernet Sauvignon ne è il primo attore e uno sfondo è affidato al Sangiovese (20%), vitigno chiantigiano per antonomasia. Questa preminenza fa dell'Etrusco un vino che guarda attraverso e oltre la chiantigianità legandosi forse ad uno dei più importanti «archetipi» dell'enosfera etrusca: il sontuoso Lucumone. Difatti, come vedremo, più vini legati ai fasti dei regnanti etruschi sono impostati su Cabernet.

Il *Re-sacerdote* a Montecchio veste arancio (il colore dell'etichetta, su cui primeggia lo stemma dell'azienda) e porta «note complesse di Pompelmo», distinguendosi da altri casi di *Etruschi,* che tendono a soluzioni estetiche un po' più fredde. Piace gustarlo immaginando, oltre i maestosi vigneti dell'azienda, il levare del sole - l'Oriente da cui l'antico popolo doveva provenire, ricco di colori e vivide suggestioni riadattate ora ad un presente locale. E a pensarci, lungo tutto il crinale che conduce da Montecchio a Castellina e ancor oltre, verso Fonterutoli con la sua necropoli, molti sono i poggioli isolati, le case-torri e casali che celano (o potrebbero celare) importanti dati archeologici. Dolci colli invasi da vigneti che ospitarono gesta, sogni e riposi di antichi signori - di cui questo *Etrusco,* tra aromi campestri e finali interminabili, evoca l'ancestrale autorevolezza.

Il riferimento letterale alla collettività etrusca ricompare con forza e pertinenza più a sud, verso il cuore dell'antica Tuscia, ad esempio in Val di Cornia con *N'Etrusco*, interessante miscela di Sangiovese-Cabernet-Merlot. Lo troviamo in vendita anche nel negozio del Parco archeologico di Populonia (la città che prenderebbe nome da *Fufluns-Phuphluns*) in una pregiata confezione-cantinetta. L'azienda produttrice è Rigoli, posta non lontano dal parco, rinomata per i suoi vini e le numerose iniziative culturali. N'Etrusco riepiloga un *bouquet* tipicamente etrusco con bagliori d'effluvi di sottobosco e frutta, la stessa che si faceva crescere nei pressi delle antiche vigne per depistare animali e insetti. Una vigna compare al centro dell'etichetta che reinterpreta, su campo nero, una pittura vascolare dionisiaca, mostrando la divinità dotata del suo *kantharos* (la coppa per bere). Elegante e morbido, riscalda il palato e l'anima - già inebriata dalla bellezza di luoghi incredibili come il golfo di Baratti.

Non lontano dalla stessa costa etrusca, a Castellina Marittima, la fattoria di Carvin produce una linea di bottiglie definita «Etrusca» e presenta vari tipi di vini (dal Bolgheri al Chianti) tra cui un ottimo Vermentino. Nel marchio della Cantina il Grappolo si fonde all'Anfora; la metafora si "addensa" al primo sorso e diviene struttura, il gusto è severo ma danzante. L'Etrusca – la declinazione al femminile risulta assai efficace - guarda ai luoghi

Bottiglia Etrusca baciata dalla luce del tramonto.

degli antenati: sentori di mango e frutti esotici, estese persistenze su orizzonti vicino-orientali…il tragitto antico di un nettare primigenio.

Spostiamoci ora ad est, verso l'Appennino e quindi l'Umbria. Per un altro vino di nome *Etrusco* fanno da testimoni gli stessi che ne hanno curato il *marketing* grafico (*Archi's*), descrivendone le caratteristiche in un breve testo reperibile su web. L'Etrusco di 3Esse, azienda agricola di Castiglione del Lago è prodotto «in modo naturale», è impostato su Sangiovese e si presenta in tre versioni: Primo rosso (giovane), Rosso e Riserva (affinato due anni in *barrique*). Per la comunicazione, gli *autori* sembrano sfruttare anzitutto l'*input* del "primo rosso", creando una linea di etichette vivaci, ben più solari della media dei vini etruschi, spesso un po' "cerimoniosi" se non "sepolcrali". Le trame di colori appaiono circoscritte dal profilo del lago Trasimeno, l'*Etrusco* è prodotto infatti nella zona ad Ovest del lago «dove si fondono le millenarie tradizioni di Umbria e Toscana» [*S.*]. Il lago vulcanico ha di per sé un simbolismo pienamente etrusco: spesso si parla di Bolsena e del vicino *Fanum Voltumnae,* ma anche il Trasimeno fu considerato un "cerchio sacro" celante un possibile ingresso agli inferi. Una leggenda racconta che lì sarebbe sprofondato Trasimeno figlio di Tirreno (primo re degli Etruschi) dopo aver sposato la ninfa Agilla. Volendo riallacciarsi alla leggenda, suggerisco che lo spirito di questo *Etrusco* possa evocare l'idea di un amore intenso e giovanile.

Anche se talvolta i creatori di etichette e i curatori della comunicazione-vitivinicola sono ricordati e persino celebrati, nel caso dei vini etruschi questo sembra capitare raramente. Sarà perché spesso e specialmente in certe zone, la scelta etruscofila è automatica, spontanea (penso ai Colli Etruschi viterbesi, alla Maremma o alla Val di Cornia) e non richiede eccessivi "protagonismi" grafici; in fondo può bastare un nome arcaico e un'immagine archeologica (come avveniva già in certe bottiglie più "antiche"). D'altro canto riproporre l'estetica etrusca con gusto, sintesi e originalità sembra tutt'altro che semplice.[3]

Scelte sofisticate si osservano nell'*Etrusco* di Col di Betto,

---

[3] Se si focalizza in modo vago nel concetto di *Etruschi*, senza un vero e proprio *punto di forza*, si rischia di favorire soluzioni generalistiche; difatti questa scelta potrebbe coinvolgere altri ambiti della vita quotidiana, oltre il vino, le bottiglie, le etichette. Il lavoro dovrebbe quindi essere improntato anzitutto sull'immaginario del rapporto tra *Etruschi e vino* e su ciò che esso comporta.

prodotto sempre in provincia di Perugia, alle falde di monte Tezio. In questo caso la *E* dell'alfabeto fenicio, di color rosso, si sovrappone all'immagine sintetica del disco di Magliano (l'antica *Heba*). Il procedimento di realizzazione vinicola è altrettanto complesso, oltreché biologico. L'uvaggio è autoctono del territorio: il Gamay del Trasimeno vendemmiato con attenta selezione, fermentazione con rimontaggi giornalieri, macerazione post fermentativa e affinamento in acciaio. Il colore rosso rubino intenso è parte integrante dell'estetica delle bottiglie che nelle etichette riprendono, con decisione, esclusivamente il nero e il rosso. Gli aromi sono «fruttati e floreali», eleganti e complessi. Al palato si presenta «morbido all'attacco, ma allo stesso tempo pieno e deciso», persistente e con finale «leggermente amarognolo» [*S.*].

È evidente come nei pressi del Trasimeno l'appellarsi direttamente agli Etruschi sia percepito come un gesto importante, da svolgersi con le dovute attenzioni per ottenere un risultato degno degli antichi signori. Tuttavia, come già detto, questi ultimi non si definivano «etruschi» e la stessa *E* ci parla di una possibilità di fascinazione più *nostra* che *loro*.

L'evocazione del *patto* etrusco si complica un po' qualora venga utilizzato uno degli autentici nomi che gli Etruschi davano a sé stessi. Nelle fonti antiche troviamo più accezioni del popolo e di chi «ne fa parte»: da *Rasna* a *Rasenna,* da *Tusci* a *Turusha*, a *Tirreni/Thyrsenoi* ecc., potremmo definirlo come un "popolo dai molti nomi" – e questo testimonia forse la sua eterogeneità, la capacità d'assimilare clan ed etnie nella "federazione" incentrata sul noto *Fanum*. Per gli abitatori del Mediterraneo orientale l'Italia fu una specie di "nuovo mondo" dove concetti contemporanei come «integrazione» e «intercultura» rivestirono forse una fiera funzionalità. Questa molteplicità di nomi e di idee può riflettersi ancor oggi nel variegato mondo dei *vini etruscofili.*

Riprendiamo proprio dal nome «Rasenna»,[4] definizione insondabile nelle sue possibili interpretazioni: «rasena» dall'egizio «rasenu», dal vedico «rasenna», dall'accadico «rasu», dal villanoviano «rhasena» (?), con significati molteplici e bizzarri (da nome delle coste siriane a quello di un fantomatico "popolo dei rasati"; da eponimo a complesso concetto devozionale, è stato scritto un po' di tutto...).

È interessante notare come alla parola *Rasenna/Rasna,* di sicuro *appeal*, corrispondano alcuni tra i vini che più hanno intrapreso, appunto, la via della contaminazione senza chiudersi, ad esempio, nell'idea che il vitigno etrusco per eccellenza fosse quello di Giove/Tinia (Sangiovese) o che si dovessero perseguire vinificazioni assolutamente tradizionali.

L'azienda La Rasenna (Ladispoli) imposta i suoi vini su più uvaggi vinificati in purezza, tra cui Vermentino, Cabernet, Moscato e *Petit Verdot.* Quest'ultimo rilancia idealmente il rapporto tra Tuscia e terre del Nord, ma anche il Moscato è raramente utilizzato in ambito "etrusco". Il terreno di produzione è assai ricco di «mineralità» e «microelementi» [*S.*], di cui le grafiche raccolgono la complessità con scelte eleganti nei colori, nelle forme (le bozze assai espressive di un cerchio o di una circonferenza) e nel raffinato logo (una mezza anfora che può ricordare anche un elmo antico, di profilo). I singoli vini prendono nome dagli uvaggi stessi e danno il massimo in abbinamenti a loro volta eterogenei, egualmente improntati sull'immaginario della costa tirrenica.

Anche la parola *Rasna*, molto diffusa nelle epigrafi, è legata

---

[4] Con questa parola rispose un uomo etrusco a Dionigi di Alicarnasso che lo fermò per strada chiedendogli «chi siete»? (I,30,3). La definizione trova conferme archeologiche nelle numerose iscrizioni riportanti «rasna», «rasnal», ecc., oltreché nell'onomastica (*Rasennius*) e toponomastica (ad es. *Rassina* presso Arezzo).

almeno ad un interessante vino contemporaneo.

Rasna si presenta come vino "da signori", bottiglia da champagne, disegno elegante – prezzo comunque abbordabile. L'Azienda è Torraccia di Chiusi, nei pressi di San Gimignano. Il vino è disponibile in tre declinazioni basate su Sangiovese (in rosato), Verdicchio o Pinot Nero. Di notevole interesse quest'ultima: il Pinot Nero, uvaggio non semplice da trattarsi, è vinificato in bianco e il colore tende dunque al ramato. Il *bouquet* è incentrato su agrumi e accompagnato da un'equilibrata acidità, una sostanziale morbidezza che non indebolisce la pungente riconoscibilità degli aromi. «Rasna»: non si può non intendere questo nome al femminile o affermarne le suggestioni attraverso il potere dell'"androginità" di certe figure etrusche. Il colore ferrigno del vino è ripreso sullo sfondo della piccola etichetta dove campeggia una maschera dallo sguardo ambiguo e trasparente; la scritta del nome si legge da destra a sinistra secondo l'antica usanza. L'amalgama estetica è ben riuscita e impone qualche domanda in più sul ruolo simbolico-funzionale di una tale bevanda.

Senz'altro vino "da aperitivo" o da convivi impostati su formaggi, frutta, pesce, il Rasna corteggia quella "regione del giorno" celebrata generalmente da anonimi prosecchi ma che può esser ora recuperata ad una sua "identità etrusca". Questo vale per le sue diverse declinazioni e ancor più per la vinificazione in bianco di un vitigno tutt'altro che toscano, ma che è guardato da certi nostri vignaiuoli come una sfida in cui metter in gioco il proprio "spirito", evocato nei nomi più antichi della nostra collettività (penso al caso di Voltumna, di cui parlerò). Qualora si richiami questo "spirito" capace di accettare e contaminarsi, fondersi e rifondare, la metafora etrusca può rivelarsi decisiva per la buona riuscita della comunicazione istituzionale/commerciale.

Sempre a proposito dei nomi di popolo, un caso speciale è quello dell'azienda *Rasenna in Tuscany* di Antria (Colli Aretini) che produce vini come *Vinum* e *Nerum*. Abbiamo già visto come il fattore etrusco possa avere un ruolo nella sintesi di un *marketing* efficace, ma raramente registreremo un tentativo di ricreazione di quello che poteva *essere* e significare il vino nell'antichità - quindi un suo fondato rinnovamento. *Rasenna in Tuscany* realizza vini «come gli antichi» secondo il cosiddetto «metodo Mondini». [*I.*]

Come testimonia il produttore stesso, a *Rasenna* si produce "all'aretina", prevalentemente con uve nere e con un 15% di uve bianche; gli uvaggi sono Sangiovese, Trebbiano, Malvasia e Albana. L'invecchiamento avviene in orci appositamente creati da artigiani cortonesi, coibentati a mano con cera e resine; per un anno e più il nettare sosta in anfore sepolte e viene estratto attraverso una cannella che fuoriesce dal terreno sottostante il terrazzamento...

L'immagine del vino o dell'acqua che sgorga dalla terra sarebbe propria dello stesso etimo di «*vinum*», come spiega Donatella Tognaccini in uno dei suoi ultimi libri sul Chianti (che di per sé significherebbe appunto «colline da cui sgorga acqua»).[5] La tradizione dell'invecchiamento in orci interrati è tipica dell'Asia Minore da cui si dice provenissero gli stessi Etruschi. Mondini stesso parla dei suoi orci interrati come di una piccola necropoli. Non a caso era usanza degli antichi costruire, in onore degli eroi, tombe dotate di particolari condotti attraverso cui versare il vino all'interno del sepolcro (esistono attestazioni di queste usanze, ad esempio, in Peloponneso). Ebbene, le suggestioni sono molteplici e vanno ben oltre l'immaginario etrusco evocato dai nomi e dalle immagini delle etichette - per altro in

[5] D. Tognaccini, *Se la vita è solo il nome che porti*, Press & Archeos, Firenze 2015, p. 93.

Antefissa con volto di Fufluns, riproduzione visibile ad Antria (Rasenna in Tuscany).

questo caso non particolarmente elaborate.

Restano davvero affascinanti questi metodi d'invecchiamento con orci interrati che come nel caso di Martville in Georgia si praticano ancora nei luoghi dove è avvenuta l'originaria domesticazione della vite. L'interramento degli orci vinari non è frequente nell'attuale Etruria, anche se negli ultimi anni è cresciuto l'interesse per l'utilizzo della terracotta. Sembra che il cotto d'Impruneta abbia proprietà isolanti tali da non richiedere, se utilizzato per conservare il vino, l'interramento; ma certamente, porre anfore sottoterra riattiva valenze metaforiche da non sottovalutare. A parte i risultati tecnici, fare-vino è anche questione di gesti: di ritualità.

Se è vero che parlare di "etruscherie" significa, ormai da secoli, parlare anche d'oltretomba e di cavità sotterranee, con la "eno-necropoli" di Antria non abbiamo contraddetto la tendenza: spesso la fantasia sembra precedere il dato storico-realistico, e la vitalità degli antichi e del loro popolo è una cosa sola con la loro stessa *prospettiva di morte*. Si tratta insomma di una vitalità archetipica.[6]

[6] James Hillman, *Il sogno e il mondo infero*, Est, Milano 1996, pp. 65 e seg.

CAPI DI POPOLO: IL LUCUMONE VESTE VIOLA

Nella *polis* etrusca il *Lucumone* (*Lauchum*) è il detentore del potere laico e religioso. La parola, curiosa e affascinante, dà forse l'idea di una superba e flessuosa continuità. Più produttori contemporanei hanno battezzato i loro vini con questo nome-concetto.

A primo acchito può sembrare incongruo che vini dai nomi eminentemente etruschi, come il *Lucumone* di Mezzetti (Cortona) siano prodotti al 100% con uvaggio Cabernet-Sauvignon, essendo gli Etruschi spesso associati, con approssimazione fantastica, a vitigni "italici" come il Sangiovese. Tuttavia proprio il Cabernet si affermerà inevitabilmente, per la sua diffusione e duttilità, tra i vitigni più rappresentativi della viticultura etruscofila. Inoltre sulla scia dell'immaginario notiamo che il Cabernet (*Kapnios* il suo antico nome), proprio come gli antichi Tirreni, proverrebbe dall'Oriente del Mediterraneo se non dall'Epiro - o almeno si è pensato questo per molto tempo.

Regale e maturo il DOC cortonese si fa strada nell'anima e nel palato in modo avvolgente, attraverso lunghe persistenze, con profumi e sentori - persino balsamici. Le indicazioni per l'abbinamento sono più sofisticate del solito: non solo carni arrosto ma anche in umido; zuppe vegetali saporite: viene in mente quella "scottiglia" che è stata idealmente ricondotta proprio agli antichi Rasna. *Lucumone* ha inoltre un'estetica calzante ed elegante, improntata sul viola con caratteri fenici-etruschi in etichetta, distinguendosi dalla tendenza "frugale" di altri vini etruschi che conosceremo.

Ma di "Lucumoni" ne esistono diversi e dalle soluzioni sorprendenti, come un IGT prodotto a Chiusi, impostato su Cabernet Franc e Merlot ma soprattutto su Petit Verdot (Azienda Ficomontanino), con etichetta piacevole e colorata);[7] ed un altro vinificato con Vermentino, da Magliano in Maremma (Azienda Mantellassi) - linea sempre elegante, caratteri allungati, colori similari: insomma, il Lucumone *veste viola*. E non possiamo non pensare alla magia intrinseca di questo colore, alle sue tonalità che ricordano da un lato le arci-figure delle cerimonie cristiane e, dall'altro, i fegati sacri analizzati dai pagani per determinare il futuro. Non è in fondo, la pelle dell'Apollo di Veio, dipinta di colore viola?

Tra tutti i lucumoni e capi di popolo non possiamo dimenticare il primo, quello che guidò la migrazione degli antichi Etruschi nell'attuale Toscana, oggi dedicatario di un ottimo vino prodotto da Mazzei: Tirreno, capo dei *Thyrsenoi*. Della migrazione si parla nella controetichetta, certo semplificando un po'. Il vino è elegante e netto, di taglio bordolese con Alicante e Petit Verdot, prodotto da vitigni rivolti al mare Tirreno, a Belguado, nei pressi della foce dell'Ombrone che compare nella cartina in etichetta.

...Ecco i cabernet che "danzano" nella "stiva" di *quelle navi* tra sentori di vaniglia, note di frutti estivi e ribes nero, echi di legni di barrique. Legni "affettuosi": travi di uno scafo onirico che attraversa un mare morbido, sapido, originario. Un mare interno e intenso, ripiegato su se stesso da onde armoniose come colli; Mare e Maremme sì, ma forse anche qualcosa del Chianti...nella struttura e nella poesia.

I principi etruschi, di cui la storia non ci ha lasciato tracce determinate, possono tornare protagonisti con fugaci apparizioni nelle etichette dei vini attuali. A Casale Marittimo (PI),

---

[7] Recentemente evolutasi in una soluzione più minimale, a campo allungato, con scritte eleganti su sfondo blu oltremare.

nell'entroterra di Cecina e verso Volterra, Pagani De Marchi produce un sangiovese «di costa» in purezza, la cui bottiglia richiama ancora i *Rasna* con l'immagine di una statua antica, secondo uno schema già consolidato: è il Principe guerriero. Vinificato con metodi tradizionali, il Principe sosta un anno in bottiglia e un secondo anno, per l'affinamento, in *tonneaux*, con un risultato complesso ed elegante [*S.*].

Ma a comandare in Etruria non erano solo lucumoni e principi. Quello che nella romanità si chiamava «pretore» qui era detto *zilath*, principale magistratura cittadina, nonché definizione di un vino di Casale Cento Corvi, della zona di Cerveteri. Parliamo di un bianco affinato in acciaio, impostato su Trebbiano e Malvasia coltivati poco sopra il livello del mare, su un terreno d'origine vulcanica; sensualità e freschezza, sensazioni olfattive marine con aromi amplificati dalla Malvasia. L'etichetta è tripartita in verticale e reca intrecci dai colori che richiamano rame e oro, generando una speciale autorevolezza. Parleremo ancora dei vini di questi produttori.

A proposito di antichi sacerdoti e collegi depositari di funzionalità religiose-divinatorie, torniamo in Chianti dove troviamo un altro "etrusco": *L'Aruspice*, vino dell'azienda Est di Belvedere (Greve). L'aruspice nel mondo etrusco è colui che interpreta il volere degli dei attraverso la lettura dei fegati ovini posti in analogia, attraverso modelli bronzei, con la volta celeste e il movimento degli astri.

Del vino di Est di Belvedere, incentrato sui sentori e sul carattere del territorio chiantigiano, si dice sia «uno dei migliori (...) oltre che rispettare i criteri del biologico» [*R.*]. Non molte su internet le tracce di queste pregiate quanto rare bottiglie; ho conosciuto l'Aruspice attraverso il vinsanto omonimo: la visione aruspicina si confà al *vino santo* in cui la traccia del divino è netta all'intuito, per il colore e per la profondità dei sentori.

Cielo e territorio, mondo umano e macrocosmo s'intersecano in un'amalgama soave. I passiti sono difatti prima *vini degli dei* poi più generalmente dei *santi* ed è stato già osservato, in una recente pubblicazione,[8] la loro accezione "paganeggiante".

[8] AA. VV., *Vinsanto come si crea come si comprende*, Quaderni di Montagliari vol. 1, Press & Archeos, Firenze 2016, pp. 39-40.

Nomi di Divinità[9]

Nella ricerca sui vini che portano nomi etruschi non potevamo non imbatterci nel dio *Fufluns*, protettore del vino, dei vignaiuoli, dei bevitori e degli antichi iniziati. Nel panorama dei vini etruschi *Fufluns* non è però tra i nomi più diffusi, forse a causa del suono che risulta complicato e un po' buffo ad orecchi contemporanei; ma credo dipenda anche dalla poca conoscenza di questa figura della nostra antichità.

Nel viterbese, in località Loiano, viene prodotto un IGT che porta precisamente il nome *Fufluns* aggiungendo in etichetta, su sfondo grigio, un antico simbolo-carattere tratto da un'iscrizione esistente nei pressi di Loiano, forse legata alla viabilità locale [*I*]. Vinificato con Trebbiano e Chardonnay presenta un colore «paglierino brillante»: il dio dell'ebbrezza non si associa solo *rosso*. È d'altra parte curioso che la stessa azienda (Olivieri) produca anche un vino di nome *Dioniso*, un «rosso granata» dal carattere più classico (mentre forse, per il nome autoctono *Fufluns*, ci si concede qualche sperimentazione in più).

L'intensa sapidità di questo bianco e gli echi delle qualità tufacee del terreno ci aiutano a stabilire piacevoli circostanze esistenziali, favorendo al contempo una sosta in quella "zona-aperitivo" già corteggiata da vini come il *Rasna*.

Più a Sud, a Cerveteri, esiste un vino che prende il nome di *Pacha*, epiteto del Bacco etrusco o secondo alcuni nome dell'iniziato ai riti enosofici. La bottiglia reca in tre etichette distinte l'immagine vascolare del dio, il nome del produttore e

[9] Per una migliore comprensione userò il carattere *corsivo* quando il nome è riferito alla divinità, e il carattere normale, con maiuscola iniziale, quando mi riferisco al vino.

quello del vino; ciò rende l'idea di una speciale complessità, e testimonia forse la ricerca di un equilibrio. L'azienda è Tenuta Tre Cancelli che si spartisce l'enosfera ceretana con Casale Cento Corvi, Cantina Cerveteri e altri produttori.

Pacha, vino di corpo e di struttura, sembra farsi strada con decisione e sincerità ritagliandosi un'identità indigena. Troviamo ancora il Sangiovese e il Merlot abbinati stavolta al Montepulciano con risultati corsivi, a tratti soavemente capricciosi. Si intuiscono le suggestioni tipiche del *terroir*, i sentori di frutta e di gelso, il tenore tufaceo ed arioso che conferma l'ascendenza infera e "vulcanica" così coerente con l'accezione dionisiaca. È evidente l'intenzione di reinterpretare il territorio e le sue docg, mostrando forse maggiore lucidità rispetto ad altri casi di vini neo-etruschi.

Un altro vino che si allaccia al Dioniso etrusco sembra essere il *Fufluna* di Poggio Rosso. Si tratta però di un riferimento indiretto: Fufluna è l'antico nome di Populonia (città dedicata appunto a *Fufluns/Phuphluns*) nei pressi della quale esiste l'azienda produttrice di questo vino «giovane ed immediato» [*S.*]. Fufluna, composto da due varietà di Cabernet, Merlot e Syrah ed evoluto in acciaio è qualificato da «grandi profumi» e può esser considerato una buona variante di un carattere «lucumonico» dei vini d'Etruria, con lineamenti assolutamente riconoscibili. Anzitutto il colore viola particolarmente intenso, a cui segue una degustazione avvolgente, struttura scaltra e vivace, pronunciati sentori di frutta e cioccolati, finale sofisticato. Da notarsi anche il tenore alcolico (ben 14°) che può favorire una nitida ebbrezza. Il divino Fufluns si esprime quindi nelle sue sembianze più ispirate e giovanili (parleremo più avanti di questa accezione del dio).

Nell'etichetta del Fufluna il nome della città/divinità è ripetuto con due diversi *fonts* (uno aggraziato, l'altro corsivo) favorendo forse una mediazione tra la frugalità della tavola e le cerimonialità degli antichi signori.

Poggio Rosso produce altri vini con nomi di divinità, declinati sobriamente a diverse colorazioni d'etichetta. Ad esser evocati sono *Feronia, Tages* e *Velthune* cioè Voltumna, garante della federazione etrusca, ricordato anche in altri luoghi della Toscana vitivinicola. Feronia è impostato sul «delicato» vitigno Viognier ed è affidato ad una divinità di stampo arcaico, essenzialmente agreste, che "protegge" una ridotta produzione di rare bottiglie. Tages (dio-bambino etrusco emerso dalla terra) è un vino più "giovane", che gioca sull'equilibrio tra Merlot e Sangiovese. Il Velthune (Voltumna) resta coerente con un'idea di giovinezza pur ponendosi in modo complesso e intrigante. Di colore rosso rubino intenso, è sobrio e corposo, caldo e tannico oltreché ricco in aromi: frutta matura ed erbe locali. Tutti i vini sono permeati dagli apostrofi di un *terroir* ricco di materiali ferrosi (*rosso*). È confermato il minimalismo della grafica delle etichette in cui compare il logo aziendale, ispirato alla decora-

Suggestioni e riflessi a Poggio Rosso.

zione di una terracotta etrusca, rappresentante una palmetta che sembra fosse diffusa nella zona [*I.*]. I vini di Poggio rosso sembrano assumere con intelligenza il paradigma etrusco tenendo la distanza dagli aspetti più sinistri/enigmatici.

Spostiamoci ora nell'Etruria del nord. A Dicomano, in Val di Sieve, il supremo *numen* Voltumna «siede offrendo le spalle all'Appennino». Anche in questo caso i viticoltori hanno sentito la necessità di rifarsi al popolo Rasna e al suo più importante rappresentante.

I terreni argillosi e la piena espressione del Sangiovese esaltano la toscanità dei vini della cooperativa *Voltumna* incontrando altresì un vitigno ospite, il Pinot Nero, sistemato cordialmente nel punto più elevato e boscoso della proprietà. A/da Voltumna si cerca evidentemente di superare gli stereotipi pur salvaguardando le tradizionali armonie toscane. *Voltumna Pinot Nero*, *Voltumna Querciolo* e *Voltumna Zeno* sono le tre declinazioni della produzione, appartenente ad un progetto collettivo e «biodinamico» dedicato alla «comprensione della comune vitalità della terra», non ha caso è "evocato" un dio della fertilità. Ma nonostante l'imperiosità del nome, il riferimento all'antico popolo non sembra protagonista della comunicazione aziendale (anche se in una bella recensione si parla delle tracce di un antico tempio nei pressi dei vigneti), né si notano ulteriori immagini etruscologiche (l'etichetta, assai raffinata, presenta solo scritte e numeri). L'aggancio agli Etruschi è quindi meditato e mediato concettualmente, né diretto né semplicemente iconico. Sono in ballo idee come *collettività*, *comunanza*, *fertilità* ma anche *eterogeneità*. «L'abitudine non aiuta l'ingegno» [*S.*], dice l'ospite: il Pinot Nero stesso che qui, appunto, «non parla francese» ma è di casa: è etrusco.

Tra le divinità amate dai produttori vinicoli è emblematico il caso di *Turan*, conosciuta come la Venere etrusca ma celebrata anche in Magna Grecia - da qui un vino aglianico: il Turan Dea prodotto nel salernitano. Nell'antica Etruria Turan è dea della bellezza così come della fertilità, fino a curiose declinazioni cultuali. È patrona della città di Vulci e protagonista (insieme ad *Uni*) nell'area sacra di Castelsecco, "mancata" acropoli d'Arezzo che ospitò un grandioso tempio, su più livelli, con tanto di odeon e mura pelasgiche.

Molti sono i vini che portano o hanno portato il nome di Turan. A Grotte di Castro (Viterbo) l'azienda Pacchiarotti produce un passito aleatico dai profumi vari e intensi con raffinata immagine/etichetta: tratti viola su sfondo nero, scritta in caratteri dorati. Anche il Turan prodotto da Torri di Torano Nuovo (da notare il toponimo che discende appunto da *Turan*), un Montepulciano d'Abruzzo «secco e armonico», reca scritte dorate su sfondo scuro (blu notte) e presenta, nell'etichetta del 2006, una pianta di vite stilizzata con rami in rilievo dorato. Vi è inoltre un Sant'Antimo, impostato su Sangiovese, sempre di nome Tùran, prodotto da Nardi di Montalcino: anche in questo caso le scritte o i fregi che attraversano la "cover" sono dorati.

Dunque, il nome della dea del *bello* e dell'*utile* ispira l'utilizzo di suggestioni auree, quasi come se le vitree bottiglie, lucenti e trasfocanti - ontologicamente legate alla sfera del femmineo -, chiedessero d'esser impreziosite da gioielli e trame dorate. Una scelta automatica, forse non originale ma certo sincretizzante, poiché unisce il *marketing* di prodotti diversi e distanti tra loro, legati quasi esclusivamente dal fattore di questa divinità.

Aplu è il nome etrusco di Apollo ed è anche il nome di un ennesimo elegante vino a base di Cabernet e Merlot, invecchiati separatamente per 18 mesi in botti di rovere. Presentato in bottiglie bordolesi da Colucci di Poggibonsi, Aplu sembra aprirsi ad un estro cosmopolita e i variegati sentori ne sono una conferma: spezie, peperone, liquirizia, menta, con il consiglio di abbinamenti dai sapori decisi, soprattutto selvaggina, ma previo prolungato invecchiamento.

Nell'etichetta compare la figura danzante del dio, incorniciata in una grafica non particolarmente «fresca», comunque elegante. Aplu è colto nel suo più intimo spirito dionisiaco a conferma di "nietzschiane compensazioni", e la degustazione sembra darne conferma; ma oggi questo vino pare difficile da reperire.

Esiste un altro *Aplu* nei pressi di Cortona: l'azienda è Villa Loggio di Cignano, vocata alle uve di varietà internazionali e produttrice di più vini dal nome antico. *Aplu* è a base di Pinot Nero; troviamo inoltre *Curtun* (Cabernet e Merlot), *Arruns* (solo Cabernet), *Tinia* (Syrah) ed altri. *Tinia* si distingue per la dolcezza, la complessità, la lunga persistenza. I sentori sono boschivi: piccoli frutti rossi, lamponi e ribes, spezie e pepe. Anche *Aplu,* corposo ed equilibrato, rimanda agli aromi del sottobosco con note di cioccolato e cuoio. Al di là del riferimento ai personaggi divini, le eleganti etichette non presentano però elementi arcaici. Ad esser protagonista è lo stemma della fattoria su sfondo nero mentre, inferiormente, a variare è solo il nome del vino. Dal punto di vista estetico il fattore etrusco sembra poco sviluppato ma resta da segnalare uno dei rari casi di vini che portano il nome del padre degli dei, Tinia, scelta sulla quale altri hanno probabilmente esitato (ne parlerò).

Tra i suddetti rari casi segnaliamo anche il Tinia di Poggetto, nel cuore del Chianti (Castellina). Un vino d'ottima personalità, ottenuto da Sangiovese e Merlot con ampi sentori terrestri, classificato «ai livelli qualitativi più alti della produzione regionale» [*S.*]. In realtà il nome è scritto *Tinìa* con un uso dell'accento non abituale - almeno nella parola etrusca - e la "componente arcaica" non pare protagonista, specialmente nell'etichetta che pure in questo caso mostra solo la scritta *Tinìa* e nessuna immagine.

La figura di Tinia non ha forse immediati riscontri iconografici, com'è al contrario per lo *Zeus* greco e questo non rende facile scegliere e pubblicare figure accattivanti quanto riconoscibili. D'altro canto, è come se le antiche *parole divine* avessero una potenza di per sé nonostante l'*impenetrabilità* dei concetti; *impenetrabilità* che, per così dire, si riflette spesso nel colore *compatto*, porpora, rubino o viola...come nel caso del suddetto Tinìa, ma anche del Fufluna ed altri.

Tornando al seguito di Fufluns, protagonista di un "parlamento" di divinità minori, spesso femminili, che compaiono nelle iconografie e presumibilmente nelle antiche ritualità, troviamo *Cautha,* quindi uno dei vini dell'azienda Le Lase nei pressi di Orte (gestita guarda caso da quattro sorelle). Cautha, sorella del Sole, figura dagli incerti contorni concettuali, dà qui il nome ad un rosso «non comune» [*I.*] e complesso. Ottenuto da Cabernet Sauvignon e Petit Verdot copre un ampio spettro olfattivo: dalla frutta rossa alle spezie dolci, dalla violetta al cuoio e caffè. Maturato in botti di rovere, d'ottima personalità, elegante e pieno, caldo e consistente, con una "severità" che si rispecchia nel colore rosso rubino *cupo* e nel gusto secco, trovan-

Orte, ingresso alla città sotterranea.

do forse analogie, appunto, con la figura di Cautha, solare ma enigmatica, legata alle ritualità dionisiache-iniziatiche. Ancora una volta la scelta del nome etrusco si associa ad un desiderio di sperimentazione e al contempo ad un'ancestrale autorevolezza.

Le Lase producono altri vini dedicati alle divinità dei Rasna; penso al *Thesan*, che poi è il nome etrusco di *Eos*, divinità dell'Aurora. Impostato su Canaiolo nero accuratamente selezionato è creato con metodologie complesse e caratterizzato da risultati ampiamente estetici. Come per il Cautha, nel caso del quale compare un'immagine solare, così le etichette di Thesan hanno un'impostazione minimale e recano il simbolo di una sorta di triskele, composto da tre forme di color viola su fondo nero.

Registriamo un certo successo del nome *Thesan* che viene assunto da un *ciliegiolo* della già conosciuta azienda Mara Olivieri e, senza la *h,* da La Biagiola di Sovana per un DOC realizzato con Sangiovese, Alicante e ancora Canaiolo nero. Pure in questo caso i produttori esprimono a pieno il *feeling* per il popolo etrusco, soprattutto con l'etichetta che riprende non solo il nome a caratteri arcaici (incisi da sinistra verso destra per una migliore leggibilità) ma anche l'ornamento "a scacchiera" con quadri rossi, azzurri e bianchi, già in uso in alcuni noti affreschi sepolcrali.

Tesan colpisce per la corposità e il tenore alcolico (14°), per la complessità del *bouquet* che scaturisce dalla macchia mediterranea, attraversa sentori balsamici sfiorando poi tabacco, liquirizia. Insomma: un altro episodio di divina autorevolezza nel cuore della Maremma toscana.

Interessante il *marketing* di altri vini della Biagiola come il *Pesna* o il *Matan:* le parole sono etrusche e corrispondono a nomi personali, a divinità locali o ad attività legate al sacro. L'azienda si distingue inoltre per una ricerca sulla vinificazione e sulla maturazione operata in grandi orci di cocciopesto, come sembra avvenisse in un antico insediamento compreso nel territorio dell'attuale azienda, dove gli scavi hanno recuperato frammenti di materiali isolanti per cisterne e vasche [*I.*][10]

Nel novero delle altre divinità titolari di vini citiamo il Mercurio etrusco, quel *Thurms* che fu protettore del piccolo *Fufluns/Dionysos*. Lo incontriamo a Torraccia di Chiusi, dove abbiamo già conosciuto un interessante vino dedicato ai Rasna (in generale). Il volto del dio compare in etichetta, su sfondo nero, tratto da una nota statua acroteriale proveniente dal tempio di Portonaccio (Veio).

Thurms s'impone al gusto, autorevole e strutturato, con tannini vellutati e ben equilibrati ad alcol e acidità [*S.*]. All'olfatto si manifestano fiori e frutti neri; il sapore è dolce e complesso con retrogusto «lungo e persistente». Un vino da signori che si allinea allo spirito dionisiaco e all'accettazione implicita del dio dell'ebbrezza in un *pantheon* olimpico, tra gli altri figli di Zeus.

Tra le divinità "minori" ed altre figure mitologiche evocate nell'enosfera d'Etruria ricordiamo il cortonese *Achelo*, vino del grande circuito Antinori e nello specifico della fattoria La Braccesca di Cortona. Proprio nei pressi di quest'importante lucumonia etrusca fu ritrovato un notevole lampadario in bronzo, con volti di divinità tra cui *Acheloo*, ricordato da Virgilio ed altri come eroe fluviale dalle forme mutevoli, a volte drago, altre toro o uomo tauroprosopo, comunque sconfitto da *Hercle* che gli staccò un corno durante una delle sue dodici fatiche. Nell'etichetta dell'omonimo vino è protagonista il volto della divinità con sorriso faunesco e, appunto, corno spezzato. Basato 100% su Syrah; di colore densamente violaceo, saporito e fruttato, con finale lungo, esprime il vigore del banchetto etrusco sospeso tra

[10] *La Nazione*, Sorano, 6 febbraio 2016.

Scena di baccanale da un sarcofago proveniente da Bolsena.

cerimonia e godimento. Il *2014* è caratterizzato da note di liquirizia, quasi tabacco (sembra che quest'ultimo sia più evidente in altre annate).

Ancora una volta la declinazione etruscofila di un vino passa per il recupero, quasi eruditico, di particolari figure, eventi, miti. Sarebbe interessante comprendere più a fondo in qual modo, la chiamata in causa di questo "saccente lucumone" e delle sue divinità, possa rendere autorevolezza a vini destinati alla grande distribuzione che di per sé rischia spesso di "livellare" la qualità del gusto.

A proposito del concetto più generale di divinità ricordiamo un importante Brunello di Montalcino battezzato *Aisna*, parola etrusca che significa «divino». Ad apparire in etichetta è proprio Fufluns, il *di-vino* per eccellenza, con tirso e coppa che viene riempita da Artemide.

Aisna corrisponde a un *blend* di Sangiovese grosso e Brunello invecchiato per 5 anni prima in botti di rovere, poi in *tonneaux* [*S.*]. Dal color rosso tendente al granata, di gradazione decisa (14°), ha sentori di fogliame e note vegetali ed esprime «una timbrica dominante che richiama classicità» [*R.*], rendendo pieno onore al suo nume titolare…

Impossibile, nonostante i mesi dedicati alla ricerca, concludere questo capitolo in modo esaustivo…altri nomi compaiono come in un rivolo d'immagini…*Eprus, Atis*, *Rhesan*, *Larthe*, *Pyrgi*, ancora *Voltumna*…tutti legati a vini che ci promettiamo d'indagare quando possibile.

Nomi di persone e nomi di luoghi

Gli Etruschi ebbero un'onomastica coerente e caratteristica, riflessa oggi in una moltitudine di toponimi e sovrapposta in certi casi a funzioni mitologiche. I nomi di persona sono stati utilizzati per definire vini assai diversi tra loro, conferendo un indubbio fascino alle bottiglie.

Sempre nella prolifica Poggio Rosso di Populonia troviamo il *Phylica*, nome di una schiava che offrì fiori e frutta agli dei (come risulta da un frammento ceramico locale). Basato su Vermentino e vinificato in bianco, conferma la predisposizione di parte dei vini d'Etruria al consumo "leggero". Fresco e assai piacevole all'olfatto, è consigliato abbinato a carne bianca o per l'aperitivo. Così come il Rasna e altri analoghi, sembra intriso dello spirito di un ipotetico "umore" etrusco: una *soavità originaria* che avrebbe animato l'aristocrazia etrusca, almeno per come la percepiamo oggi.

Veniamo dunque al *Velia*, un Bolgheri di Guado al Melo (Castagneto Carducci), di cui si apprezza la sintesi tra Sangiovese e altri uvaggi, con *bouquet* intenso e puntuale, e tannini ben percepibili.

L'immagine di pittura vascolare in etichetta (forse dal fascino ellenico più che italico) rappresenta una soluzione efficace ma non troppo elaborata; le potenzialità di questo vino sono convogliate da una comunicazione schietta ed efficace – e pure al gusto si esprime una relativa "durezza". Questo Velia sembra suggerirci l'autorevolezza *materiale* delle donne etrusche, più che qualche tipo di evanescente sensualità (come in altri casi). Queste si distinsero in ruoli chiave della società (fatto impensabile in altri popoli antichi), furono capaci di influenzare la collettività e,

Ritratto di Velia Spurinna, nobildonna tarquiniense.

a distanza di secoli, la toponomastica.[11]

Ma chi/cos'è precisamente «Velia»? certo la zona di produzione è etrusca e *Velia* era uno dei nomi femminili più diffusi nel tardo periodo etrusco. Velia è inoltre il nome di una *polis* del nord della Magna Grecia dove i Rasna ebbero un fiorente emporio; ed è al contempo un nome di persona che in epoca recente ebbe una certa diffusione, specialmente in Toscana. Si pensi al romanzo di Bruno Cicognani *La Velia* e il suo vincente personaggio, definito «vivace e frizzante». In breve, Velia è il nome di altri luoghi, alture, *clan* e personaggi come la protagonista, assolutamente etrusca, del *Fantasma del Giglio* di Fabio Fornaciari.

Velia è anche il nome di un vino rosato laziale (Grotta Baiocco) e di un altro bianco umbro (Ponziani): il riferimento agli Etruschi sembra poco sviluppato ma in un certo senso, con la vinificazione in bianco, ci avviciniamo allo spirito della femminilità etrusca. Altresì il "maschile" di *Velia* è probabilmente *Vel* e veniamo dunque a uno dei casi più curiosi tra quelli incontrati nella ricerca.

[11] Penso ad esempio a *Ramta* e al caso di *Monterantuli*, in Val di Greve.

Prodotto da Poggio Gagliardo (Montescudaio), Vel Aules ha tutte le caratteristiche per affascinare già ben prima di giungere a naso e palato. La vinificazione si basa su procedimenti tratti da manoscritti del XVIII sec., fatto che lo rende di per sé «biodinamico»; la scelta dei vitigni è tutt'altro che banale: Malvasia nera (autoctona del volterrano) e Colorino; il nome è quello del protagonista di un romanzo d'ambientazione etrusca (*Il segreto etrusco* di Giuliana Boldrini); inoltre l'etichetta presenta, pur nella sua semplicità, quel piglio "artistico" che rese famose altre etichette di altri grandi vini affidate a caparbi pittori.

La degustazione si promette ricca di sorprese. Chi si aspettava un vino «antico» è rimasto sorpreso per la vitalità, la corposità, per i profumi e gli aromi fruttati e per il color rubino impenetrabile. Secondo alcuni, Vel Aules è capace di reggere il confronto con i più sorprendenti vini sudamericani o australiani restando comunque allacciato alla tradizione toscana e locale.

In questo senso si può parlare forse di una moderata «contaminazione»: come se l'assorbimento del paradigma arcaico ispirasse, là dove si cerca di superare le soluzioni semplicistiche della commerciabilità, un coraggio ed un estro ulteriori. Non a caso, nel romanzo della Boldrini, *Vel Aules* viaggiò da Populonia a Marsiglia, dove i Rasenna stabilirono realmente una base commerciale. Come vale per altri vini che abbiamo conosciuto, l'«eroe etrusco» guarda al mondo intero, sperimenta metodi e vitigni in libertà e sfida, a suo modo, le grandi tradizioni enologiche.

Fu forse un avventuriero anche *Pesna*? È questo il nome di un vino de La Biagiola di Sovana che sembra sia stato diffuso anche come personale. Abbiamo già accennato al Pesna per la bella estetica dell'etichetta: sfondo docile, vagamente siderale, verde e celeste; una goccia in primo piano e sopra il nome in carattere arcaico. Il vitigno è il Vermentino, vinificato con preziosa cautela, lieviti indigeni, dodici mesi in *tonneux* per un risultato intenso, selvatico, persistente e ancora una volta «agrumato» [*S.*] (penso al Vermentino de l'Etrusca o a L'Etrusco di Montecchio). Ma è da notare, in questo caso, la preziosità delle bottiglie che sono prodotte in serie numerate a mano: quando si scomoda l'eroe etrusco gli si offre una qualche *unicità*, si arricchisce il corredo con sentori, note, gesti; in altre parole, in un certo qual modo, si *sacrifica*.

Un eroe etrusco realmente esistito e ricordato dalle fonti è Mastarna, a cui è dedicato un vino della Tenuta Tre Cancelli di Cerveteri. Secondo una tradizione tramandata dagli affreschi della Tomba François (Vulci), riconducibile a fatti verificatesi nel VI sec. a. C., il guerriero *Macstrna* liberò *Caile Vipenas*, prigioniero dei romani, per poi partecipare alla conquista di Roma stessa. Dal testo di un discorso tenuto dall'imperatore Claudio nel I sec. veniamo a sapere che Mastarna fu eletto re della città con il nome di Servio Tullio, dopo aver ucciso *Aule*, fratello di *Caile*, con cui era entrato in disaccordo dopo la morte del primo.

Quello che colpisce della vicenda di Mastarna è l'impressionante catena di omicidi descritti nei suddetti affreschi, attinente al conflitto tra vulcensi e romani. Un vino che porta il nome di un personaggio del genere, e che lo fa riferendo direttamente all'etimo etrusco, dovrebbe testimoniare una certa tensione e poderosità. Così, da tale prospettiva, può stupire che il vino in questione non sia un sanguigno rosso ma un pregiato vino bianco.

Gli echi degli scontri tra gli eroi del Lazio antichissimo lasciano solo qualche lontano riverbero attraverso il Procanico, unico vitigno del Mastarna. Cerchiamo di riconoscere questi segnali intermittenti, nella pienezza e sapidità, favorite dai terreni lavici della zona; nella decisa armonia, immersa nell'inquietante silenzio dei dintorni della necropoli ceretana. Questo vino sembra ricordarci che il vendicatore Mastarna era destinato a diventare Re, e tale fu: ecco allora i riflessi dorati, la pregnanza dei primi sorsi, il gradevole *bouquet*, le note quasi balsamiche. Una raffinatezza che non contraddice, anzi avvolge l'acida ispirazione dei Rasenna.

Tra i vini etruschi "maschili" sospesi tra onomastica e toponomastica segnaliamo anche *Manero*, il cui nome non sembra essere etrusco (indica di quando in quando una persona, o un luogo – il *maniero*, da *manere*, rimanere) ma appare tuttavia scritto, in etichetta, in caratteri etruschi. L'etichetta reca inoltre l'immagine di una maschera arcaica stilizzata che ricorda figure di antefisse e acroteri dei templi arcaici. Prodotto da Tenute del Cerro (Bologna) presso la fattoria del Cerro (Montepulciano), Manero consiste in un tessuto improntato su Sangiovese e un 20% di Merlot. I profumi sono quelli "istituzionali" del *Clan* etrusco per eccellenza (che diede i nomi a *Chianti* come a *Chianciano*): more e frutti di bosco a cui si aggiunge una suggestione di vaniglia. Morbido, tende forse verso il secco.

L'azienda ha realizzato altri vini il cui nome compare scritto in caratteri etruschi, come un *Brunello* o un *Nobile*. È chiaro si tratti di un rimando a un'identità territoriale ed è forse nel Manero che il "contatto" è riuscito a pieno, proprio per il gusto che traccia un vettore con i grandi binari del Sangiovese e che rievoca, appunto, la continuità del *Clan*.

Come abbiamo visto i nomi di persone e quelli di luoghi tendono a sovrapporsi, soprattutto a causa delle logiche della toponomastica prediale della romanità. Un luogo ha quasi sempre un suo *nomen* guardiano, ed ogni nome/entità ha una sua *residenza*.

Sul nome *Chianti* esistono più interpretazioni. Secondo alcuni deriverebbe da *Clante*, termine utilizzato ancora nel medioevo per indicare l'affluvio dell'attuale torrente Massellone (Gaiole) e legato ad una più antica radice etrusca, interpretata in modi diversi ma certo riconducibile al popolo dei Rasenna, alle sue sorgenti, alle sue ninfe e divinità boschive. Ma stranamente, proprio nel cuore del Chianti storico, sembra sia raro incontrare il nome *Clante* associato ad un vino. Nessuna azienda vinicola, che io sappia, presenta (almeno su web) una linea di bottiglie che assuma questa parola (salvo definizioni generali di tipologie legate al Consorzio).

Certo il Chianti, da un punto di vista geografico quanto esistenziale, è concetto vario ed espanso; possiamo dunque ricordare il *Clante* dell'azienda Borghetto di Montefiridoldi (San Casciano) impostato esclusivamente su Sangiovese, invecchiato in rovere e commerciato in piccole, preziose quantità. Nella presentazione di questo vino si accenna infatti alle colline boscose abitate dagli Etruschi del Chianti - anche se in realtà il toponimo fece la sua comparsa molto più a sud.

Forse, l'utilizzo del concetto *Clante* in ambito vitivinicolo pare più spiccatamente "etrusco" in casi esterni al Chianti. Su tutti quello dell'azienda Pomaio, posta sui monti aretini, che produce un *Clante* e anche un *Porsenna* (nome del più famoso lucumone etrusco). Il primo, dalla trama «elegante e potente» è basato su Merlot; il secondo, che «impressiona per l'ampio spettro olfattivo», [*R.*] è orientato sul solo Sangiovese. Semplici ed eleganti le grafiche delle etichette dove campeggiano le parole-concetto: tra esse «clante», non ha certo bisogno di ulteriori effetti visivi data la sua affascinante profondità e le sue recondite connessioni che allargano gli orizzonti all'antichità mediterranea e persino mesopotamica.[12] Se è vero che il concetto-toponimo di *Chianti* tende a espandersi - almeno in senso enologico, quanto turistico

[12] Penso ancora alle suggestive interpretazioni di Giovanni Semerano, *Il Popolo che sconfisse la morte*, Mondadori, Milano 2006.

-, possiamo forse tollerare questa "dilatazione" tenendo conto della portata della radice originaria.

A proposito di nomi di luoghi, spostiamoci dunque più a Sud, dove i clan etruschi fondarono Chiusi, ricco crocevia della Toscana arcaica il cui nome etrusco fu *Camars,* che è pure un interessante vino dell'azienda San Gregorio.

Sofisticato incontro di Barbera e Colorino vinificati separatamente in piccole *barrique*, sussurra tutta una gamma di *legni* che dialogono con i sentori dei frutti spontanei. Franchi lamponi, sospetti di vaniglia/liquirizia, tendenziosa persistenza, morbida personalità...il palato coccolato da un finale pastoso. Si torna a osservare il bicchiere controluce ed ecco, a conferma, ancora il *viola*: riflessi di un porpora atemporàto, ironico lucumone o brioso vescovo che sia. Colpisce per la gradazione la quale, di fatto, ispira ebbrezza dunque poesia.

Sempre nell'aretino, a Marcena, un'azienda che recupera il nome *Aritima* (che ha a che fare con l'antico nome di Arezzo - *Aritie, Aritim* – ed è anche la definizione etrusca di Artemide), produce alcuni vini biologici tra cui un rosso, un bianco e un *Chianti.* Non semplici da trovarsi in commercio (almeno per quanto riguarda il web), le bottiglie recano come logo l'immagine sintetizzata di uno scudo etrusco, con tanto di fori per gli occhi (se ho ben interpretato) e alcuni caratteri in filigrana. Una scelta interessante che evoca un'essenzialità quasi araldica: l'energia di pungenti valenze identitarie.

Tra i vini "mitici" dell'aretino spicca altresì il Cuna di Podere Santa Felicita, in Casentino. In questo caso il nome può essere ricondotto al concetto di luogo delimitato: colle, ambiente, camera (*s-cuna*), ma anche punta di terra tra due fiumi (*cuneus*), pelaghetto o *laguna.* In etichetta la parola compare scritta in caratteri etruschi, da leggersi correttamente da destra verso sinistra.

L'*imprinting* appenninico qualificato da boschi e paesaggi asperrimi sembra essere protagonista del sottofondo di questo vino, «intenso e scuro, fatto di richiami a erbe aromatiche e a sentori di sottobosco e spezie» [*R.*]. Frutto del lavoro scientifico e concettuale Federico Staderini, Cuna è impostato su Pinot Nero, vitigno che predilige una certa altitudine (a *Cuna* siamo a 500 m.) ed è da considerarsi pienamente biodinamico: vendemmia manuale, fermentazione spontanea con lieviti indigeni in tini e barrique usate, imbottigliamento senza alcuna filtrazione. Il risultato è «rustico e intrigante ma allo stesso tempo armonioso» e «di una profondità disarmante» [*R.*]. Tra i sentori, tenui ed eleganti, si riconosco frutti e spezie, erbe boschive e persino riverberi d'incenso.

Staderini è andato oltre nella sperimentazione con un vino dal nome "Sempremai Sorte", per il quale è stato recuperato l'uvaggio Abrostine, secondo alcuni il vitigno originario dell'Etruria appenninica, già chimera di altri viticoltori e archeologi. Il risultato è acido e terrigno, elettrico e «magnetico»...il rapporto diretto e sanguigno con il territorio e con certi luoghi o «posti» speciali, pare essenziale alla creazione di vini del genere.

Tra i toponimi-*enonimi* etruschi più originali spicca quello di *Hebo*, dell'azienda Petra di Suvereto. Il nome di questo vino dal gusto intenso e complesso è quello di un antico insediamento etrusco [*S.*].[13] Ma la parola avrebbe pure assonanze celtiche, oltre che definire alcune località del Medioriente e dell'Africa orientale.

Il nome è stato scelto per evocare le origini etrusche – in questo caso della Val di Cornia, le colline ferrose, quel particolare *terroir* e i relativi profumi. Dichiarando il legame con la

[13] Ci si riferisce probabilmente a quello di Magliano in Maremma (*Heba).*

terra, con i suoi guardiani ed i suoi confini, si determina la comunione tra possidente e posseduto, tra *etrusco* e territorio. In Hebo l'uvaggio di riferimento è ancora una volta il Cabernet Sauvignon, prediletto da molti vini che, nel carpire lo sfuggente ideale etrusco, cercano al contempo una strada più sontuosa e articolata. Ma troviamo anche il Merlot (40%), fidato accompagnatore del vitigno "selvaggio"; e il Sangiovese, alfiere di Toscana (10%). L'invecchiamento prevede più passaggi alternando vasche d'acciaio, fusti e *barrique*.

In una recensione si legge che Hebo «o lo si ama o lo si odia. (...) non è un vino dalle mezze misure». Gli si riconosce insomma un estro che lo distingue da altri vini etruschi, con il suo «*bouquet* aromatico che spazia dalle note di frutti a bacca rossa, sentori erbacei terrosi e di sottobosco, spezie dolci come la vaniglia, ma anche sentori terziari di cuoio e pelle» [*S.*]; sorprendente il finale, piacevolmente sapido. La forza di questo vino sta forse nel tentativo di penetrare la pienezza della sua terra in modo elegante e completo: un repertorio sensoriale che ha qualcosa della varietà onirica, e che gode d'una ispirazione *infera*. Analogamente la cantina di Petra (tra le più belle di Toscana, rappresentata proprio nell'etichetta di Hebo) penetra nella parete rocciosa della montagna attraverso «un percorso misterioso (...) un cordone ombelicale che lega alla terra e madre». Una terra/madre etrusche - toscane e mediterranee allo stesso tempo.

L'antico nome di Magliano ha ispirato un vino che si chiama altresì *Heba*, un Morellino di Scansano prodotto da Fattoria di Magliano. Il vitigno di Giove torna protagonista (accompagnato da uno sfondo di Syrah, 5%) rilanciando con «semplicità ed eleganza» [*S.*] l'immagine emblematica della campagna toscana ai cui sentori tipici (ciliegia ed altri frutti) si aggiungono suggestioni di geranio e rosa.

Sangiovese in purezza o quasi, affinato in tradizionali vasi di cemento, non può non rivelarsi equilibrato e «impeccabile», oltreché asciutto, fresco e con «tannini arrotondati», tanto che nella stessa recensione istituzionale il *sommelier* lo definisce «ottimo compagno della tavola»; un compagno eminentemente etrusco.

Felsina rappresenta un altro caso di coincidenza tra toponimo e nome di vino. La parola deriverebbe dalla divinità Velzna, o potrebbe indicare «terre fertili», o ancora «luoghi fortificati». È l'antico nome di Bologna che secondo Plinio il Vecchio fu fondata dagli Etruschi nel VI sec. a. C.; ma il toponimo definisce altre località o il loro nome-radice originario (oltre a Orvieto, penso alla stessa «Bolsena», probabile riflesso di un antico teonimo).

Una località di nome *Felsina* si trova appunto nei pressi di Castelnuovo Berardenga: qui, a circa 400 m. d'altitudine, si produce un Chianti di carattere basato al 100% su Sangiovese, dagli aromi speziati e boschivi. Anche se nel "*marketing*" del Felsina gli Etruschi non sembrano esser protagonisti, è chiaro che l'eleganza del toponimo conferisca alle bottiglie un fascino speciale: tanto che nell'etichetta poc'altro deve risaltare oltre al nome e al piccolo disegno di un «luogo fortificato». Potere della "magia" delle parole etrusche, capaci d'elevare gli elementi del quotidiano (gli oggetti, i luoghi...) al livello del sacro, e viceversa.

Ma vini, luoghi e reciproci nomi possono legarsi agli Etruschi in modi diversi, anche se non utilizzano parole originarie della lingua dei Rasna.

Il riferimento generico a una *terra* etrusca è diffuso nella Maremma laziale, nel caso di più aziende, anche se non sempre si applica al singolo vino. In tal senso ricordiamo Vulci Terre Etrusche, di produzione biologica, che produce vini con Procanico, Chardonnay, Cabernet ed altri vitigni tra cui il "raro"

Viognier; ed anche l'Etruscaia di Tarquinia, i cui vini presentano in etichetta l'immagine di un oplita con una serie di variazioni grafiche. Quest'azienda si distingue per l'approccio scientifico, con un progetto sperimentale, in collaborazione con l'Università di Pisa, sui cloni di uva e sull'analisi della loro specificità [*S.*] .

Altre aziende il cui nome si lega a un territorio dichiaratamente etrusco e a quello delle singole dodecapoli, sono comparsi e compariranno nelle selezioni di Divino Etrusco, evento enogastronomico che si svolge a Tarquinia (e non solo), nel cuore di quella Tuscia che per prima, forse, produsse fiaschi e bottiglie che dichiaravano a chiare lettere il loro essere etrusche.

La tendenza ad appellarsi alla *terra* etrusca non poteva mancare nella zona della Costa Etrusca/Valdicornia. Ad esempio il Terratico di *Bibbona*, della linea Poggio degli Etruschi, imbottigliato da Ferrari (Cecina), prodotto con Merlot, Sangiovese, e affinato in *barrique*, ci avvolge con un gusto capace di sintonizzarsi su valenze terrestri e sulle influenze di un toponimo chiaramente etrusco (*Vibius/Bibius*). Mi ha colpito in particolare il testo nella controetichetta che recita: «*...ripercorrere la storia del vino significa scrivere la storia dell'uomo. Sapere e Sapore, Cultura e Coltura, vita e vite (...) al tempo degli Etruschi non esistevano confini tra il vino, la spiritualità e la vita quotidiana (...) erano diffuse pratiche in onore di Bacco...*». Chi ha pubblicato questo testo aveva, come dire, le idee molto chiare. E guarda caso esso ricompare (?) nei *depliant* del viticoltore etrusco-carismatico per eccellenza: il Mondini di Antria che abbiamo già conosciuto.

Anche la *Valle delle Stelle*, un *biologico* di Brancatelli, sempre in Valdicornia, sembra parlarci di un luogo d'ascendenza etrusca. La scritta sulle etichette ha un carattere arcaistico ed è dorata: sono come parole sussurate da Turan, in attesa del Dio, nelle notti populoniensi.

Ecco il bicchiere di un violaceo intenso: ed il gusto ci sorprende. Livree lignee distese su corpi di tradizioni toscane, *barriques* e pietra-panchina, rumori di bosco e cipressi sfilacciati dal vento marino... Ottima sintesi di Cabernet, trattiene al centro qualcosa d'importante che s'impone al primo sorso: un cuore scuro, etrusco, avanza persistente, rotola nel palato e raccoglie stelle come briciole; le distende, le struttura in ordini caotici; Orione ed altri "lucumoni"...cosmici e cosmopoliti, velluti del Nord più caldo, incanti originari, confidenze boreali... - dopo aver conosciuto già molti vini, viene spontaneo poetizzare e il fattore etrusco sembra chiaro all'intuito…

Mare etrusco nei pressi di Baratti-Populonia.

Concludiamo il nostro "riepilogo" con un concetto essenziale ad ogni località quanto necessario a tutto il mondo etrusco: quello del *confine*. Avremmo voluto saperne di più su un'azienda di Sassofortino (Grosseto) che ha prodotto un vino dal nome importante, *Tular*, parola etrusca che indicava le pietre di confine delle proprietà private, così come delle città/lucumonie. L'indagine su questo vino, il cui sito-web ufficiale pare oggi estinto, può basarsi, al momento, solo su qualche memoria in *google-immagini*; frammenti suggestivi come la fotografia di una damigiana piena di tappi di sughero posta in un bosco...come fosse appunto una specie di *tular*.

Le pietre confinarie erano intrise di una sacralità che il vino può rendere accessibile attraverso le etichette, sorta di *lapidi vivaci*, forti di segnali grafici e dei loro relativi simbolismi. I Tular erano confitti, attivati e onorati con procedure religiose in momenti speciali; la loro presenza era tanto rispettata che molti sono rimasti al loro posto sino al medioevo, e oltre. Lo stesso senso di rispetto è evocato, con un pizzico d'ironia, da quella damigiana silvestre proprio perché, in quanto vino, essa gode già di una dignità "sacrale"...

Tular non è solo un vino ma pure una birra (una delle pochissime "birre etrusche") prodotta dall'azienda umbra Goretti. È disponibile sia in versione *chiara* che *rossa* e si presenta con posata eleganza. Il logo consiste in tre linee, pennellate o graffite, che ricordano proprio i tratti confinari. E, tornando ai vini, anche Goretti lavora con Pinot, stavolta *Grigio*, ed anche con Chardonnay, e c'è pure un vinsanto dal nome *Sanctus*.

Oggetti, giochi, animali

In un'indagine sulle produzioni vinicole che si rifanno all'immaginario etrusco non potevamo non imbatterci in un vino che portasse il nome del cottabo, gioco greco ed etrusco largamente praticato durante i banchetti, consistente nel lanciare fiotti di vino in ciotole sospese lungo un asta.

Ora ci riferiamo al *Kottabos* di Casale Cento Corvi. Prodotto nelle colline nei pressi di Cerveteri dove i Rasenna ebbero la loro più grande necropoli, Kottabos sembra occupare una posizione speciale nella "geografia psichica" dei vini etruscofili.

Se fosse realmente possibile intuire, nell'ambito dell'enosfera etrusca, tendenze "lucumoniche" (generalmente vini rossi, maschili, maestosi, avvolgenti, accompagnati da immagini penetranti) o maggiormente "eteree" (bianchi o comunque tendenti al secco; femminei, leggeri, dolci, ricchi di sentori e colorazioni in etichetta), potremmo dire che il Kottabos, come *concept*, si pone al centro del rapporto.[14] E lo troviamo, forse non a caso, sia "bianco" (con Chardonnay e Trebbiano), sia "rosso"; noi ora riferiamo al secondo.

Dal profumo dolce, intenso e fruttato, è impostato su Merlot e affinato in rovere, percorrendo un sentiero che lo rende «morbido, caldo» con «note dolci» e «raffinata eleganza» [*S.*]; ma le origini tufacee del terreno da un lato e l'immancabile presenza del Sangiovese dall'altro (30%) sembrano acconciare queste armonie a un equilibrio e una struttura tipicamente "etruschi". Nella completezza di questa determinazione Kottabos può evo-

---

[14] Lo stesso strumento del Cottabo può esser certamente inteso come il simbolo di un equilibrio assiale, declinato a una dimensione ludica.

care l'atmosfera del gioco di cui porta il nome, nel quale Etera e Lucumone, ebbri a fine pasto, collaboravano al lancio del vino che l'uomo faceva zampillare dal *Kilyx* o dal *Kantharos* (come si osserva in una nota illustrazione di Anatolio Scifoni).

Cento Corvi, già produttore dello Zilath, ha appunto ripreso anche il concetto di *Kantharos*, nome della coppa per bere che sostituì la più arcaica *Kylix*.

Anche il Kantharos si distingue per la pertinenza e la capacità di sintesi. Nato dal connubio tra Merlot e Syrah, è invecchiato un anno in bottiglia. Il risultato è «selvatico e quasi primitivo» [*S.*], incentrato su profumi di macchia mediterranea con suggestioni sapide e saline: si intuisce la vicinanza del mare, a pochi metri dalla cantina di produzione. Affascinante l'etichetta con astrazioni su sfondo nero e nome dorato.

Quando si parla di oggetti che danno nomi a vini si riferisce generalmente a strumenti impiegati nel consumo e nella produzione di vini ed altre essenze. Aska è un altro di questi casi.

I vini di Banfi fanno da tempo la loro parte nelle tavole di Toscana, grazie alla riconoscibilità e ad una capillare distribuzione. Penso ad esempio al *Col di Sasso* col suo "colle infuocato" in etichetta; meno conosciuto e più pregiato è Aska, vino di taglio bordolese che Banfi ha iniziato a produrre dopo aver acquistato un terreno nel territorio di Bolgheri. Gli Etruschi usavano la parola *aska* per definire i contenitori di vino e olio, profumi e nettari, dunque «le emozioni e i sentimenti che gli uomini ricevevano da Cauha e Semla, il sole e la luna, rappresentati da due sfere luminose che scandiscono, con il giorno e la notte, il trascorrere del tempo» [*S.*]. Nell'etichetta si notano due piccoli cerchi (sole e luna) perfettamente integrati al posto dei trattini delle *A* di *aska*; le lettere sono distese e aggraziate da una moderata vettorializzazione; lo sfondo nero pone in risalto questa semplice ma raffinata composizione.

I vitigni di riferimento sono appunto Cabernet Sauvignon e Cabernet Franc; già abbiamo visto quanti vini etruschi si rifacciano al Cabernet che è al contempo l'uvaggio prediletto di Bolgheri. Il gusto è «intenso, morbido», il colore rosso rubino con riflessi violacei, mentre il *bouquet* comprende spezie e frutti rossi, «legno tostato», caffè. La struttura è dunque complessa, armoniosa e equilibrata, con la capacità di adeguarsi ai diversi abbinamenti mostrando sempre qualche sfumatura qualitativa inedita. Aska, da vino DOC ed "etrusco" quale è rammenta, nel complesso della sua comunicazione, l'incedere di altri "lucumoni" viniferi - risultando forse più strategico e sfaccettato.

Anche *stamnos* è il nome di un contenitore: stavolta si parla del vaso dal corpo globulare con cui Greci ed Etruschi potevano servire le bevande. Un vino che porta questo nome è prodotto in terra falisca, a Fabrica di Roma, da Menicocci. Stamnos è un raro caso di Merlot in purezza prodotto con uve rigorosamente

Ingresso infero in una tomba a Populonia (Buca delle Fate).

biologiche, lieviti autoctoni, fermentazione naturale e in un'ottica dichiaratamente *vegan*. [*S.*] Questa accezione è molto rara nel caso dei vini etruschi, tendenzialmente rossi e associati alla cacciagione. Nell'etichetta compare lo stesso Fufluns, con il figlio Oinopio, tratto appunto dalla pittura di uno *stamnos*; come in altri casi d'evocazione del dio riappare il colore rosso-viola-granato, oltre che sentori di liquirizia. Si registra una «non semplice» riconoscibilità varietale data «la territorialità e la totale assenza di Solfiti» [*S.*]. Stamnos fa propria la "scelta etruscofila" (anche con un altro vino, *Rhesan*), percorrendo con coraggio, con un *rosso*, un'assoluta biodiversità.

L'enosfera etrusca continua a stupire per la qualità e per la ricercatezza concettuale dei singoli episodi, tanto che proseguendo la ricerca non possiamo non riaffacciarci a Torraccia di Chiusi dove, a proposito di vini ispirati a nomi di oggetti, incontriamo un vino che ha il nome...di una fibula etrusca. La parola è legata al ritrovamento di questo reperto del VII sec. a. C., a forma di luna, nei pressi di una vigna e significherebbe appunto «oggetto».

Nome o semplice concetto, comunque da leggersi al femminile, *Cver* corrisponde a un vino di color giallo paglierino, quasi dorato, con profumi floreali, speziati e minerali e un complesso *bouquet* dal quale «emergono sensazioni di ginestra, pesca gialla, ananas, papaya, vaniglia, pepe bianco, pietra focaia, eucalipto, salvia»... [*S*]. Un vino morbido e al contempo imponente che conferma le capacità dell'azienda produttrice, una autentica "garanzia etrusca".

Sul simbolismo degli animali è invece impostata una linea di vini di Urlari, in zona Bolgheri. Etichette davvero raffinate con rappresentazioni di un Cavallo (Urlari), un Cinghiale (Pervale) e un Lupo (L'Urlo), ispirati dal repertorio delle figure arcaiche. Gli sfondi sono colorati rosso/seppia/azzurro e presentano incisioni in oro che descrivono, nella parte superiore, speciali configurazioni astronomiche. È dichiarato il rimando agli Etruschi e al rapporto sacrale con le fasi lunari: «come accade alla luna stessa durante le sue fasi, quello che ora è vuoto si riempirà nuovamente, se abbiamo la pazienza di aspettare». [*S.*] Si legge inoltre che gli Etruschi «importarono la vite dall'Oriente» e, se riusciamo a crederci oggi…figuriamoci cosa potessero pensarne gli Etruschi stessi.

La sintesi estetica di Urlari sembra alquanto riuscita ed i vini stessi rivelano un crogiuolo di elementi, ingredienti e metodologie cosmopolite. Vitigni di varietà bordolesi impiantate con sangiovese, vinificazione alla francese, enologi con esperienze in varie parti del mondo...un vino tutto da scoprire anche se la "componente etrusca" non coinvolge il *nomen* che, nonostante la buona assonanza, ha tutt'altra origine.

Un altro vino di cui voglio solo accennare l'esistenza è omonimo e abbinato ad un oggetto particolare: un *libro,* ovviamente odierno. Si tratta del romanzo *I segreti della via etrusca* di G. Bracci e M. Parlanti. Prodotto da Cantine Guidi (Poggibonsi), il blend è basato su Sangiovese e Canaiolo nero, tra i più antichi vitigni già esistenti al tempo degli Etruschi, come si legge in quarta di copertina...*pardon,* nella contro-etichetta.

Degustazione e Lettura, Etruschi e nuovi fermenti editoriali…non possiamo non riconoscere agli autori e curatori (Itaca Edizioni) un notevole brio: durante la presentazione fiorentina del libro si sono fatti addirittura scortare da due soldati-figuranti etruschi...

DALL'ARCHEOLOGIA DELLE ETICHETTE AL MERCATO INTERNAZIONALE

Possiamo dire che il fenomeno dei vini etruscofili si sia affermato, in buona parte, di pari passo con il turismo archeologico, inteso anche e soprattutto come risorsa identitaria per le popolazioni locali. Bottiglie con la dicitura «vino etrusco» o «colli etruschi» sono rintracciabili dalla seconda metà dello scorso secolo nell'ambito di una produzione locale, create su iniziativa delle singole aziende, mentre alcune DOC che riferiscono a territori etruschi sono nate prevalentemente nella metà dagli anni '90.

Non è facile scoprire quali siano stati i primissimi vini in commercio, anche perché la definizione di *vini di Tuscia* o *vini d'Etruria* è già implicita nei trattati d'enologia toscana del secondo rinascimento, quando le *etruscherie* rappresentavano un importante riferimento estetico per una riflessione identitaria (questo forse già dai tempi di Cosimo I).

Ma un commercio notevole di vini chiaramente intitolati agli Etruschi non poteva aver inizio se non con il diffondersi delle *bottiglie* e delle loro etichette, a discapito di fiaschi e terrecotte: allora i nomi e le immagini di riferimento presero un'importanza e una dimensione maggiore investendo le necessità della comunicazione commerciale. Non avendo a disposizione cataloghi specifici dedicati al passato (un passato precedente tra l'altro alla nascita delle *denominazioni* e talvolta degli stessi consorzi, quindi probabilmente spesso "irregolare" o "perdutamente" locale), resta possibile affidarsi al rarefatto mondo del collezionismo di bottiglie ed etichette antiche. Qui, in cantine dismesse o mercatini d'antiquariato, e ovviamente negli annunci su web, capita d'incontrare qualche vino etrusco. Com'è il caso ad esempio delle "antiche" confezioni dei Colli Etruschi Viterbesi o di qualche vino di Maremma.

A tal proposito, credo sia importante notare un interessante effetto psicologico-sensoriale. Ci si trova a osservare, su certe vecchie bottiglie polverose, altrettanto vetuste etichette recanti, nel caso dei vini d'Etruria, immagini arcaiche e quindi ulteriormente "antiche". Si genera quindi un *feed-back* cognitivo molto particolare. Questo può indicare forse la direzione di una ricerca sul senso del rapporto tra vino e sacro, attraverso un ideale scavalcamento dei limiti cronologici.

Inoltre le immagini delle etichette evocano un'intuizione quasi mai portata alla coscienza ma immediatamente chiara all'anima di qualsiasi etruscofilo: una ricerca che, passando per la scoperta di antiche bottiglie recanti immagini ulteriormente vetuste (soprattutto riproduzioni di pitture e incisioni arcaiche), sintetizzi la possibilità di una *collezione* appunto *pseudo-archeologica*. Siamo, insomma, nel pieno di un sentimento romantico-etruschero.

A evocare queste emozioni sono ad esempio le etichette di un vino de La Parrina, azienda considerata storica e rappresentativa del territorio etrusco di Orbetello. Il vino è un *Etrusco* progenitore dei numerosi vini che portano oggi questo nome e che abbiamo già considerato. L'immagine in etichetta, consunta non meno dell'originario dipinto arcaico, è quella di una donazione di una coppa a una divinità...il gesto prediletto da Fufluns. L'*Etrusco* compare nelle liste dei collezionisti ed è ben quotato da *Wine-searcher*, riconoscendo un valore che credo sia anzitutto estetico.

Le etichette, autentici specchi del "sogno enosofico", rendono misteriosamente accessibile un certo "sesto senso". Attraverso il bicchiere, realmente bevuto o solo immaginato, pare d'accedere ad antichi vincoli identitari. Riusciamo a tornare indietro di qualche decade ed è subito evidente quanto sia diverso, rispetto ad oggi, il modo di *comunicare*. Penso all'etichetta di un *Rosso Etrusco* di Montefiascone. Come avviene per altre vecchie bottiglie non è dichiarato il vitigno, ma possiamo immaginare la "sintesi" tra più uvaggi...comunque sia il risultato fu un vino secco, pensato probabilmente per un consumo schietto; da qui l'estetica incisiva con questo rosso pompeiano e le scritte i cui caratteri paiono più greci che etruschi... Forse, solo qualche decade fa, in Etruria "vino etrusco" significava anzitutto "condivisione", tavolate di popolo, "craterizzazioni" in amicizia.

Del resto, indipendentemente dal nome del vino (che può alludere a concetti ulteriori) è tutt'altro che estinta la tendenza a porre in etichetta immagini tratte dal repertorio etrusco. I casi in Maremma (laziale e non) sono molteplici e talentuosi: Fontana Morella (Cerveteri) con l'affresco dei suonatori, Vignamurata (Sorano) con l'uomo del sarcofago degli Sposi, Le Pupille (Scansano) con una tipica scena di banchetto, Poppetta di Lotti (Sutri) con l'immagine di Vel Saties e del servo Arnza (dalla Tomba François)...ma probabilmente di esempi potremmo trovarne a decine, raffinati così come più "grezzi", sondando il passato recente della viticultura d'Etruria.

Sembra comunque che i primi casi di vini in commercio che attinsero dall'immaginario etrusco siano generalmente riferibili al territorio maremmano mentre nell'Etruria centro-settentrionale si preferiva guardare, ad esempio, all'idea di Chianti nelle sue varianti e tradizioni locali. Certo, con qualche eccezione.

Si procede a volte con scoperte casuali e apparizioni improvvise, in alcuni casi capaci di portarci molto lontano perché se da un lato le radici del fenomeno sono locali e insondabili (non esiste una storiografia definibile tale), dall'altro le diffusioni etruscofile sono state sorprendenti, sino all'attualità internazionale. Ad esempio, durante una pausa in un bar del Chianti notai, esposta in cima a un mobile, una bottiglia di grande formato recante l'immagine di noti affreschi etruschi... l'etichetta presentava la dicitura «Chianti Docg», un fatto di per sé interessante poiché come abbiamo visto solo in tempi recenti è stato affermato con decisione il rapporto tra Chianti ed Etruschi. Ma da dove veniva quella vecchia bottiglia? Candoni - De Zan, azienda veneta, presenta tutt'oggi un'intera linea di vini improntati sugli Etruschi.

Le attuali grafiche dei vini "etruschi" di Candoni sono molto diverse rispetto a quella vecchia "edizione". Presentano colori tutt'altro che banali, vivaci ma rarefatti negli sfondi, mentre le figure in primo piano hanno un contorno marcato e puntuale. La fantasia etrusca sembra esser stata orientata e digerita attraverso un *marketing* maturo, comunque sognante, attento senz'altro alle necessità del mercato americano dove Candoni sembra esser forse protagonista del "riferimento etrusco". Tra i vitigni prescelti, oltre a Merlot, Syrah e Moscato spicca, ancora una volta, il Pinot in due versioni, Grigio e Noir. I Pinot si affermano dunque, al di là di ogni aspettativa iniziale, tra i vitigni prediletti dal paradigma etrusco.

Con l'esempio di Candoni, che copre un'idea di *radici* sino all'ampiezza del commercio internazionale si conclude – per il momento – il nostro *excursus* nel mondo delle bottiglie etrusche. Resto consapevole che un compendio esaustivo fosse improbabile e mi scuso, al contempo, se qualche importante vino è rimasto escluso.[15]

Dalla conoscenza dei singoli vini sono emerse alcune caratteristiche del fenomeno, con interessanti riverberi estetici quanto

---

[15] Ma esistono le *ristampe* e le *nuove edizioni* per rimediare, si attendono quindi eventuali suggerimenti e ulteriori informazioni.

forse antropologici, commerciali quanto ideali.

I vini etruschi tendono, più che altri, all'identificazione con figure mitologiche e divine ma anche con personaggi, funzioni e concetti della società arcaica. Questo è un tratto inconscio e più o meno costante del fenomeno, anche se nel tempo il rapporto si è evoluto, raffinato, si è fatto meno *letterale*. Si è passati da un semplice riferimento etimologico o iconografico ad una mimesi più complessa, con vini *d'alto livello* che sviluppano e poi scremano il materiale estetico-immaginario attraverso scelte consapevoli e sofisticate, sia nell'apparenza, sia nel gusto, nell'estetico come nell'*estatico*. In pratica, come abbiamo visto, un tempo bastava *offrire* un'immagine di pittura vascolare in etichetta scrivendo «vino etrusco»; oggi questa semplicità, pur custodendo una sua sensibile autenticità, è solo una scelta tra le tante o una "prima tappa" che può esser superata.

Indipendentemente dal *livello* dell'elaborazione, in questa enosfera etrusca emergono alcune tendenze e definizioni, come quelle relative ai riferimenti maschili e femminei che produrrebbero scelte diverse nella caratterizzazione del prodotto. Nel primo emisfero si distinguono le figure del Principe/Lucumone e del Dio/Demone nelle loro diverse polarizzazioni; cioè la mascolinità positiva, regale, suadente ed il suo negativo più incalcolabile e sepolcrale. Analogamente la donna, colta nel suo emisfero, risulta la protettrice del convivio, con sofisticate capacità che vanno dalla bontà della Schiava alle sottigliezze dell'Etera. Lucumoni e Matrone, come detto, restano forse gli alfieri di questo contesto in cui talvolta il vino bianco tende ad esser *femmineo* e quello rosso *mascolino*. Del resto nelle etichette, al di là del grado di complessità, le tendenze sono soprattutto per il bianco (con scritte nere) o il nero (con scritte nere o rosse) e gli sfondi solo raramente presentano sfumature colorate (come avviene per Candoni o Biagiola).

Il Cabernet, al di là delle tipologie e delle diverse vinificazioni (in bianco e rosso) resta il vitigno che meglio rappresenta l'Etruria, per le qualità organolettiche, la larga diffusione, la qualità espressa nell'incontro con *terroir* molto diversi. Ma quando ci si pone in modo più complesso, e i rapporti in gioco vengono ricondotti ad una ricercata minimalità, si manifestano vitigni come i Pinot e non pochi uvaggi locali, recuperi di per sé a volte quasi "archeologici" (penso a Rasenna in Tuscany, Poggio Gagliardo, Terre di Vulci o altri).

Con sorpresa non resta in primissimo piano, ma comunque importante, il Sangiovese, presente soprattutto nei territori da sempre legati a questo vitigno (su tutti il Chianti). Sono inoltre coinvolti, con una certa frequenza, Verdicchio, Syrah, Chardonnay e altri.

Dal database al mito

Ad una percezione basata soprattutto su internet, i vini palesemente legati agli Etruschi risultano essere sull'ordine del centinaio o non molto di più. Per molti di essi il rimando è palesemente dichiarato, con nomi di divinità e di altri concetti o con immagini antiche; in altri casi il fattore etrusco è meno pronunciato se non rarefatto, ma comunque riconoscibile.

Da un punto di vista, diciamo, statistico, non si può certo mettere sullo stesso piano un vino/azienda il cui *marketing* è improntato dichiaratamente sui *Rasna*, ed un altro che riflette semplicemente, se non inconsapevolmente, la sua appartenenza al territorio, assumendo quindi il toponimo del luogo, della zona o del relativo del consorzio.[16] Ma è proprio considerando

[16] Talvolta l'azienda stessa ha un nome etrusco; se i nomi dei singoli vini di quell'azienda sono etruschi, sarà da valutare ogni singolo vino come un caso a sé. Se invece questi

anche a questi vini "spontaneamente etruschi" che si raggiunge una quantità minima - comunque in costante aggiornamento – per poter osare qualche considerazione generale, pur necessariamente prudente e approssimativa.

I vini etruscofili tendono al rosso - rubino, granata o "violaceo" che sia - in circa il 60% dei casi, confermando un'accezione legata all'immaginario di Toscana e al consumo di selvaggina; ma esistono, come detto, molti bianchi e qualche rosato, spesso molto raffinati. Sia bianchi che rossi possono avere caratteristiche gustative complesse con sentori assai vari, legati alla macchia e non solo; e non sembrano particolarmente "granitici" né "monoformi".

Ad oggi le province maggiormente interessate sono toscane ma, forse con qualche sorpresa, Grosseto e la sua Maremma non s'impongono per prime all'attenzione. La maggioranza dei casi la troviamo nell'aretino e nel livornese (venti casi o poco meno per ciascuna provincia), nel viterbese (una dozzina circa), nella provincia di Siena (circa dieci) e in provincia di Roma (almeno una decina di vini, soprattutto grazie all'area di Cerveteri). Le attuali province di Grosseto e Firenze contano un numero di casi analogo – che non pare superiore alla decina. Limitata ma significativa la presenza in Umbria (concentrata specialmente nei pressi del Trasimeno); assenti o quasi i vini etruscofili a Lucca e Pistoia, mentre si registra qualche iniziativa nella provincia di Pisa. Esistono poi, come abbiamo visto, casi sporadici fuori dall'Etruria.

Interessante notare che i vini il cui grado alcolico è maggiore della media, attestandosi intorno ai 14° o più, corrispondono a

---

rimandano semplicemente agli uvaggi o a nomi non etruschi, o comunque non hanno rimandi etruschi salvo il logo aziendale, può esser preferibile considerare l'azienda come un unico caso o "voce". Per queste ed altre ragioni (come l'esistenza di casi di vini estinti) risulta difficile quantificare l'effettiva presenza di "vini etruschi" nelle singole province.

nomi di divinità (*Tesan, Fufluna, Turms, Aisna* e altri). Del resto una buona parte di vini etruschi presenta nomi mitologici (almeno 25%-30%) anche se la maggioranza, come detto, è forse legata ai luoghi che hanno conservato una definizione etrusca. È significativo, per l'identità regionale, che almeno una quindicina di vini siano dedicati ad antichi nomi di popolo o capi di popolo: i vari Etruschi, Rasenna, Lucumoni. Infine, l'altitudine media dei vitigni pare sia poco elevata, salvo casi eccezionali e legati a uvaggi speciali. Si resta insomma "al di qua dell'Appennino".

Alcuni vini tra quelli incontrati - circa una decina o poco meno - paiono "estinti", cioè fuori commercio: la scelta etruscofila è forse più delicata di quel che si possa pensare inizialmente, e può deludere per più ragioni.

Sorvolando sugli aspetti pre-statistici, è chiaro che questa indagine, pur muovendo da basi materiali, punta a datità ulteriori. Per approfondire vantaggi e svantaggi della predilezione per le immagini etrusche dovremo assumere elementi non concernenti alla produzione contemporanea, guardando anche al passato storico-archeologico.

Tappi di vini Etruschi.

D'altronde il *fattore-rasna*, o se vogliamo il *demone* del vino etrusco non sembra fermarsi ai vini palesemente dedicati agli Etruschi, quanto sopravvivere e manifestarsi più sottilmente al di là dei riferimenti letterali (nomi, immagini) come se la sua più perfetta evoluzione fosse la *mimesi,* componente "di profondità" dei vini più autenticamente legati al territorio e alla sua antica storia.

Si pensi – ma è solo uno spunto - ad un vino famosissimo, composto da un sapiente trattamento di Cabernet e Sangiovese, dove sembra essersi nascosto quel lucumone-padre degli dei il cui nome raramente è stato accolto con forza: *Tinia-(...).* È forse una semplice suggestione, ma può aiutarci a intuire la sfuggevolezza del *numen.*

Quale siano oggi le recondite ragioni di questa "entità", da dove provenga la sua forza, e in che senso siano orientati i *propositi* che si sovrappongono ad essa, sono affari complessi: c'è da scendere nel profondo della bevuta, dal gusto iniziale alle oscure conseguenze dell'ebbrezza, dalle sue funzionalità primigenie agli atteggiamenti e modi d'essere contemporanei. E c'è forse da ripartire dall'*inizio.*

Etichetta di un vino di Montefiascone, XIX sec.

Abbreviazioni (cap. I)

I. = informazione ricavata direttamente da un produttore / direttore / enologo / vignaiuolo o proprietario di vigneto.

R. = da una recensione facilmente reperibile su internet.

S. = dalla scheda stessa del vino, pubblicata sul sito dell'azienda o altra pubblicazione istituzionale.

Nota di redazione

Quando si invia in stampa un libro come questo è sempre troppo presto. L'argomento è vasto e indefinito e i tempi editoriali, come spesso avviene, non hanno permesso di perfezionare gran ché.

Chiunque volesse segnalare errori, apportare informazioni utili per altri approfondimenti o comunicare elementi che ci sono sfuggiti, non esiti a contattarci. L'autore è a disposizione, attraverso i riferimenti della casa editrice, per migliorare le prossime edizioni.

Si specifica inoltre che questa ricerca non è legata a nessun produttore vinicolo in particolare, né a enti o altre istituzioni e che l'autore si è mosso nella totale indipendenza, di pensiero ed economica.

II

gli Antichi

LA *VISIONE* DELLE ORIGINI

La coltivazione della vite ebbe principio in Oriente e giunse nei *paesi del tramonto* attraverso l'interscambio culturale caratterizzante il Mediterraneo dei primi secoli dei metalli.[17] Originari probabilmente del nord del Caucaso, gli antenati dei nostri vitigni si sarebbero diffusi in Europa e in alcune zone dell'Asia prima dell'ultima glaciazione. La glaciazione stessa li avrebbe "spinti" verso l'Europa del Sud.

Che anche gli Etruschi siano «giunti dall'Oriente», o che questa origine sia stata solo sognata o ipotizzata, resta lecito, e forse necessario, indagare nella direzione dell'Est: soprattutto se consideriamo il rapporto esistenziale che intercorre tra i popoli e i loro vini. E, per la natura della nostra ricerca, non possiamo eludere una fantasia che sembra serpeggiare sullo sfondo: è forse ipotizzabile che gli uvaggi diffusi tra gli Etruschi siano giunti con essi dall'Oriente? Come abbiamo visto, questa suggestione affiora tutt'oggi qua e là, anche nell'ambito della comunicazione dei produttori vinicoli. Ma quasi certamente si tratta anzitutto di una fantasia.

La comparsa della lavorazione della vite può essere il risultato di «puntiformi fenomeni di paradomesticazione»,[18] che condussero alla transizione da una *vitis silvestris* ad una *vitis vinifera*. Sono ormai provate la coltivazione e il consumo d'uva, in Etruria, già nell'età del Bronzo. Grandi quantità di vinaccioli, di viti selvatiche come di viti coltivate sono stati ritrovati a San Loren-

[17] Si veda anche Barbara Lucrezia Paganelli, *cit.*, pp. 240-241.

[18] Attilio Scienza, *Attualità dello studio della vite selvatica, una comune area di ricerca tra archeologia, antropologia e biologia*, in *Origini della Viticultura, Atti del Convegno*, Podere Forte, Castiglione d'Orcia 25 giugno 2010, p. 6.

zo a Greve (Bronzo Medio), a Chiusi (Bronzo Finale), a Stagno nei pressi di Livorno (inizio età del Ferro). Già i villanoviani cominciarono a interessarsi alle piante di vite (*labrusca*) che crescevano spontaneamente ai lati dei campi, stendendosi generalmente sul terreno tra rifiuti e macchia; selezionarono e potarono meticolosamente le migliori, organizzandole rudimentalmente intorno a tutori vivi e producendo l'antesignano di quello che sarà definito *temetum*.

Successe che nel frattempo - come provano gli esami del DNA e le ricerche sui principali vitigni storici - giungessero dalla Magna Grecia le *vitis vinifere,* e che i maestri vignaiuoli etruschi riuscissero a incrociarle con le varietà silvestri. L'unione fu "baciata" e custodita dal clima e dal terreno di Tuscia. Comparsero così i predecessori dei nostri Canaiolo, Malvasia, Sangiovese, testimoniati oggi dai vinaccioli emersi in numerosi siti archeologici, da quello di Gran Carro presso Bolsena (IX sec. a. C.) a quello di Poggio Bacherina presso Chianciano (III-II sec. a. C.).

Così, al tempo di Numa Pompilio, si passava dalla «protoviticultura» etrusca a vigneti strutturati e le *lambruscaie* parevano già un ricordo "romantico", relegate con i loro *rari grappoli* in *luoghi silvestri*, così come le osserva il pastore Mopso nelle Bucoliche di Virgilio (V) – sebbene le due forme di coltura, *vinum* e *arbustum*, sarebbero coesistite all'incirca per tutta l'antichità.

Quella sopra esposta può esser considerata, in sintesi, la realtà storica (la nutrita bibliografia sull'argomento è già disponibile attraverso una preliminare ricerca su web). Vi è poi, come detto, la leggenda che rivendica una sua importanza antropologica, dal punto di vista dei simboli e dei valori identitari.

Nel momento in cui gli Etruschi assunsero (o mantennero) usanze e soluzioni estetiche dal riverbero orientale, riconoscendosi in esse nonché in una ritualità che aveva nel *uinu/vinum* uno dei suoi più importanti *medium*, risultò forse scontato intravedere la *visione* di un'origine altrettanto orientale del vino, della pianta e della *legna* stessa di cui erano fatte le sacre statue di Tinia/Fufluns (di cui parleremo). Evidenziare quindi la dimensione esistenziale su cui è possibile stendere e, per quel che è oggi possibile, comprendere la tradizione del legame con l'Oriente - ovvero, su tutto, la leggenda della migrazione di Tirreno -, può rivelarsi utile alla comprensione di una parte delle usanze, degli utilizzi, dei modi di essere legati al vino degli Etruschi. Così com'è di probabile importanza sottolineare la *necessità* stessa dell'idea di un *fattore orientale*, che più volte, nella storia d'Occidente, si è manifestata, specialmente nelle fasi in cui i popoli-nazioni e le loro classi dominanti, oltrepassati gli apici di civiltà o di potere, si avviavano ad un'inevitabile decadenza.

Da un certo punto di vista antropologico le due più antiche tesi sulle origini degli Etruschi, quella di Erodoto che riferendo ad una tradizione anatolica li vuole provenienti dalla Lidia (*Storie* I,94; ripresa poi in chiave "pelasgica" da Anticlide di Atene ed Ellanico di Lesbo) e quella di Dionigi di Alicarnasso che li vuole autoctoni (*Antichità Romane,* I,26-30) restano compatibili, così come lo sono le diverse istanze che compongono la complessità umana; esse rappresenterebbero momenti diversi d'affermazione di una dignità storica e, secondariamente, possibili realtà cronologiche.

La leggenda riportata da Erodoto avrebbe potuto trovare corrispondenza nelle credenze di parte del popolo etrusco, impostato senz'altro su diverse etnie – quindi legato a matrici italiche - ma indubbiamente affascinato, almeno in un secondo momento, dall'Oriente come ci testimonia l'arte e non solo. D'altronde già in quel tempo le tradizioni, le memorie collettive e la loro ipotetica persistenza potevano alimentare il fondamento della leggenda delle origini e dare conferma ai suoi possibili tratti cronologici.

Su questa prospettiva, non dovremmo quindi immaginare una o più grandi migrazioni: in molte occasioni i popoli anatolici e vicino-orientali ebbero contatti, rifugi, empori nella penisola italica, rilasciando conseguentemente miti, tradizioni, fantasie; in una parola: *storie* (ciò di cui è *fatta* idealmente *specie umana*). Ma entriamo maggiormente nel dettaglio di questi indizi per evidenziare un'intermittente sequenza di appigli e conferme ad un'*idea*[19] che tende, oggi come allora, a diffondersi; e tenendo altresì presente come l'intuizione di un *fattore orientale* possa coincidere con un fascino di cui il vino stesso è portatore (come già abbiamo visto e come approfondiremo, per quanto concerne i vini contemporanei).

Ipotizzando che i signori etruschi intuissero un loro legame con l'Oriente, essi avrebbero potuto porre l'arrivo dei loro antenati in un'epoca miti-storica se non mitologica. Nella tensione di una corrispondenza storiografica, quest'epoca può coincidere con gli ultimi secoli del II millennio a.C. In quel tempo comparvero nel Mediterraneo popolazioni belligeranti ricordate dalle fonti egizie e ittite, giunte forse anche in Etruria: i "Popoli del Mare", *Tursha* o *Turusha*, di matrice indoeuropea, non anatolici ma caucasici, forse portatori di elementi propri della successiva cultura villanoviana; oppure poteva trattarsi di popolazioni dell'Arzawa, anatoliche, non indoeuropee e potenziali portatrici dell'essenza fonetica della lingua etrusca. Secondo alcuni, *Turusha* è collegabile proprio a *Thyrsenoi*, nome con cui i greci definivano gli Etruschi.

Sempre a proposito di una fase assai remota, la quantità di ceramica d'origine egeo-anatolica trovata negli strati arcaici

[19] Da un punto di vista analitico-esistenziale, le contemporanee ricerche sulla genetica degli etruschi potrebbero essere considerate un caso *ontico* d'affermazione di questa idea.

Dionigi di Alicarnasso in un'incisione tratta dal Codice Ambrosiano.

costieri dell'Etruria, ben differenziata dalla cultura protovillanoviana e villanoviana, potrebbe essere indizio, più che di una semplice relazione commerciale, dello stabilirsi di piccoli gruppi giunti a più riprese, già dal XIII sec. a. C., tanto da divenire abbastanza numerosi rispetto alle popolazioni locali. In perenne ricerca di minerali e altre materie prime questi *clan* si fusero con le popolazioni apportando le loro conoscenze nell'estrazione del ferro, ben reperibile in Etruria, e nella produzione bronzistica dove s'intuisce un contributo tutt'altro che sporadico.

Andando avanti con i secoli non mancano appigli altresì concreti alla "visione" di un legame con l'Oriente, e a possibili riproiezioni leggendarie. È probabile che dopo le turbolente vicende succedutesi verso la fine del VIII secolo a.C. con le campagne militari assire di Salmanassar V e di Sargon II in Siria e in Palestina, avvenissero fughe di tecnica specializzata verso l'Italia Tirrenica. Successivamente, nel 690 a.C., quando il regno di Frigia fu travolto da un'invasione di Cimmeri, si verificarono altre migrazioni.

Entriamo quindi in una fase pienamente storica.

Sembra che lo sviluppo nel VI secolo a.C. a Vulci di un'importante scuola di ceramica detta "pontica" fosse dovuta a profughi *ioni,* che si erano sottratti alla conquista persiana (del 546 a.C). Inoltre, un'iscrizione neo-babilonese in caratteri cuneiformi con l'indicazione del proprietario (*Nabû-Iddin*), incisa su un frammento di vaso bronzeo proveniente da un contesto funerario a Faleri (metà del VII secolo a.C.), spinge a ipotizzare che uomini del vicino Oriente si fossero integrati nella vita locale, restando legati a usi e culti della loro terra originaria.

Sempre a proposito dei rapporti con l'Oriente, veniamo a qualche aspetto linguistico e culturale. Non dimentichiamo che la parola/nome considerato originale degli Etruschi, *Rasna* o *Rasenna*, si ritrova sotto varie forme nei dialetti dell'Asia Minore e persino nelle lingue vediche. Pure il nome "ellenizzato" degli Etruschi, *Tyrrhenoi* o *Tyrsenoi,* avrebbe un'origine anatolica e sembra derivi dalla parola *Tyrra.* Oggi sappiamo di una località della Lidia che fu chiamata precisamente *Tyrra*: sia pure con molte difficoltà siamo tentati a trovare un rapporto tra i vocaboli etruschi e quelli dell'Asia Minore, attribuendo importanza a queste analogie.

Ad esempio le radici *Tar* e *tarch*, che si ritrovano spesso nella lingua etrusca (basti ricordare *Tarchon*, fratello o figlio di Tirreno e fondatore di Tarquinia) compaiono in nomi vicino-orientali di divinità e prìncipi, vedi il dio della tempesta *Tarhunda* o il re d'Arzawa, *Tarhundaradu*, vissuto al tempo di Akhenaton (XIV sec. a. C.). Ma la ricerca di similitudini del genere ha già portato più ricercatori[20] assai lontano, forse anche con qualche eccesso.

Il sistema politico degli Etruschi era quello delle città-stato, governate da monarchie con un *lucumone* o re scelto fra i primogeniti delle famiglie nobili che, come già abbiamo visto, esercitava il potere amministrativo, giudiziario, militare, nonché l'autorità sacerdotale con i sui complessi rituali. Tutto questo è paragonabile a quanto avveniva già molto tempo prima presso i Sumeri (III millennio a.C.), dove i re erano al contempo rappresentanti religiosi; inoltre la parola etrusca *Laukum*, tradotta in *Lucumone* e attribuita oggi a vini e altri prodotti che si rifanno all'immaginario etrusco, sarebbe un composto di due termini, uno sumero, l'altro accadico. Lo stesso etimo *Rasenna* (altrettanto utilizzato, come abbiamo visto, in ambito vinicolo contemporaneo) significherebbe *dominio di signori*: in accadico *rasum* corrisponde proprio a *capo*.

Aprendo un po' alla fantasia, potremmo sostenere che pure nell'architettura dei templi etruschi sia intuibile una matrice orientale. In un rilievo assiro del VIII secolo proveniente da *Khorsabad* è raffigurato un edificio che sembra proprio - certo usando un po' d'immaginazione - il predecessore dell'architettura dei templi dell'Etruria. Ma già sapevamo come i templi etruschi, con i loro grandi basamenti, i tetti aggraziati, la moltitudine di colorazioni, fossero caratterizzati da un che di "esotico" (almeno a uno sguardo contemporaneo).

Restando in ambito religioso, è opinione condivisibile che la sacra "scienza" di predire il futuro osservando il fegato delle

---

[20] Si rimanda in particolare agli studi di Giovanni Semerano, ma anche a molte altre pubblicazioni o testi su web che offrono varie interpretazioni e ipotesi.

vittime o il volo degli uccelli giungesse agli Etruschi dal mondo mesopotamico. Gli antichi testi babilonesi mostrano regole d'interpretazione minuziosissime e a volte simili a quelle etrusche. L'interpretazione dei fulmini costituisce una delle forme più caratteristiche della scienza dei *Rasna*: come in Mesopotamia, presso i Caldei, anche in Italia si esaminavano i fulmini per i vaticini.

L'interpretazione dipendeva dalla regione celeste da cui scaturivano e dal luogo in cui s'infrangevano; dal periodo dell'anno, mese e giorno in cui l'evento si verificava; dall'aspetto, colore ed effetto. Furono gli Etruschi dunque ad introdurre in Europa queste pratiche religiose e quindi trasmetterle ai romani, scavalcando pressoché totalmente il mondo ellenico.

Anche la mitologia etrusca presentò, con il passare del tempo, interessanti variazioni rispetto a quella greca con episodi che coinvolgono l'origine orientale. Pensiamo al mito di Giasone. Di ritorno dalla Colchide l'eroe avrebbe fatto tappa sull'isola di Lemno, innamorandosi della figlia di Myrina, regina delle Amazzoni; l'avvenimento venne rappresentato dagli Etruschi già

La Stele di Lemno, recante iscrizioni in caratteri compatibili con quelli etruschi.

nel VII secolo, su un'olpe ceretana in bucchero. A Lemno è attestata un'iscrizione compatibile con la scrittura etrusca, che ha ispirato molte ricerche sulla lingua detta "lemnia" e sulle sue origini.

Nel ripercorrere tutte queste corrispondenze, analogie, semplici suggestioni ma anche indizi tangibili sul rapporto tra Etruria e Oriente; e soprattutto nel farlo dopo aver conosciuto gli attuali vini etruschi, i loro nomi e la loro emozionalità, non si può non aver l'impressione che questo fattore diciamo "esotico" sia una componente integrante del mistero etrusco e che giochi un ruolo essenziale - quanto spesso sottaciuto - nel *marketing* delle bottiglie come nell'*ebbrezza* della bevuta.

Possiamo forse ipotizzare che nell'incedere del *Lucumone* e nella sua suadente autorevolezza, come nell'eleganza di *Turan* e nel suo splendore dorato, rivivessero il lusso, il profumo, la suggestione delle corti vicino-orientali; che i loro *aska* contenessero spezie e aromi di provenienza anatolica, medioorientale, indiana; che l'attenzione per i riflessi e la qualità del loro vino fosse la stessa riservata ai fulmini; che i padri di *Vel* e di *Velia* li avessero cresciuti a suon di leggende e fiabe avvolgenti; che le lune e le stelle di quelle *notti* fossero le stesse a cui ululavano le bestie di *Urlari*, e che sotto quei firmamenti si stendessero le città di *Heba, Camars, Kainua*...in attesa di nuovi levanti favoriti da *Thesan*.

Questa *miniera* di elementi storico-leggendari e arcaico-sensoriali ci suggerisce che *quei* lucumoni e i loro popoli, colti nell'antica, straordinaria realtà a cui aspirano e che in parte ricreano i vini "etruscofili", avessero già buone ragioni per immaginare un qualche legame con l'Oriente - nonostante la convinzione più diffusa in antichità fosse certo, realisticamente, quella dell'autoctonia[21]; che un tale contesto potesse caratterizzarsi in sfaccettature a noi ignote, come il racconto epopeico della fusione di più genti e clan (*Turusha, Rasna, "villanoviani"* e altri) diretto da un condottiero *Rasenna*; ipotesi capaci di dare un senso alla prismaticità della nazione etrusca in un'epoca già prossima alla decadenza, in cui l'unità politica doveva essere affermata con decisione in funzione del contrasto all'avanzata di Roma; ma chiaramente, anche dopo: quando urgeva mantenere una propria identità collettiva nonostante la sottomissione ai potenti vicini.

L'origine orientale della *sostanza* del popolo etrusco, realisticamente improbabile e forse impossibile nella sua primarietà

Scorcio su una degustazione in un giardino romantico-etrusco.

[21] Alcuni miti come quello di *Taiges*, *puer* nato dalla terra, oltre che approfondite interpretazioni dell'etimo *Rasna* ed altri indizi, fanno pensare che gli Etruschi si identificassero con il territorio italico, fosse solo per la necessità di rivendicarne il diritto estendendo la loro egemonia. A tal proposito e sull'autocoscienza etrusca si veda Luciana Aigner Foresti, *Gli Etruschi e la loro autocoscienza*, in *Autocoscienza e rappresentazione dei popoli nell'antichità*, Vita e Pensiero, Milano 1992, p. 95 e seg.

antropologica, è dunque una realtà esistenziale, capace di agire, nella misura in cui è stata *dichiarata*, accordandosi da un lato a dati storici frammentari, dall'altro a una necessità identitaria legittima e attualmente riconoscibile – nonché alimentata dagli stessi vini contemporanei; perché in questi *Etruschi sognanti* c'è molto anche di nostro, e ciò non deve essere inteso, dal punto di vista della presente ricerca – che è una ricerca sui vini e sull'ebbrezza - come un "inghippo", quanto come un'occasione per l'intuito.

Così, percorrendo questa ipotesi e tornando attraverso essa alle proporzioni della storia, ci viene spontaneo valutarne la forza nel contesto derivato dall'*urto* storico che stava già per infrangersi sul popolo *Rasna*. Nel corso del V-IV secolo, i riferimenti nazionali degli Etruschi si fecero forse più rigidi e dovettero condensarsi in una rinnovata autorevolezza. Mentre prima i Greci poi i Romani premevano a Sud e i Celti guadagnavano terreno più a Nord, i *Rasna* cominciarono forse a guardare, sognanti, ad Est, al luogo dove sorge il sole e dove, non a caso, molti altri videro condensarsi la fantasia delle origini.[22]

Il vino, con i simbolismi della sua produzione e il sofisticato intreccio di fantasie - di matrice personale e di base collettiva - che l'ebbrezza innesca nei bevitori, rappresenta il *cratere* su cui questa *visione originaria* poteva prendere energia e, oggi come allora, attecchire. Noi stessi, insieme al *vignaiuolo etrusco,* bevendo un vino di Tuscia immaginiamo e rivendichiamo, avvinghiati nei nostri meandri personali, una qualche *origine* e un qualche *Oriente* – inteso anche, più in generale, come *orientamento* a qualche futuribile soluzione: rinascita o rivoluzione che sia.

[22] Questo aspetto non è semplice da delinearsi e gli approfondimenti sono da rimandarsi ad altra sede. Diciamo che, con premesse e sviluppi diversi, un fenomeno analogo avrebbe investito, ad esempio, le cultualità dei Longobardi come quelle dei crociati del XI secolo; le ricerche di certi nazionalisti del XX come la spiritualità di certi rivoluzionari del XIX.

...Ecco che gli Etruschi, o almeno questi *etruschi re-immaginati*, ritengono di vivere in quella che, da tempi immemorabili, era detta *terra del tramonto* e ribaltata questa immagine, guardando a sé stessi, alle loro tradizioni ma voltando le spalle al loro imminente *tramonto*, riscoprono o adottano il mito di Tirreno, del suo *sogno* e quindi *del suo vino,* condotto in quelle antiche navi insieme alle *sacre piante* e ai loro semi.

Ed ecco che li vediamo ostentare tra i loro "*sintomi*" quelli che potevano risultare caratteri identitari. Vediamo affermarsi la moda del triclinio, già visto nelle immagini vicino-orientali[23] e nei vasi greci; dei giochi intorno al vino, dei colori esagerati e del sesso facile. Vediamo prender campo la più mediterranea e al contempo orientale delle loro divinità: Dioniso, dai caratteri spiccatamente ellenici ma che ricalcavano un *numen* più antico, *Fufluns*, testimone di un'epoca in cui il *dio del vino* fu protagonista del *pantheon* appenninico e, conseguentemente, protettore appunto dei vignaiuoli (come vedremo). Ma a conferma di tutto questo, e di più, lo stesso *Fufluns* sarebbe rapportabile ad un sostrato più antico e dal riverbero mediterraneo: comunque, guarda caso, connesso all'Oriente...

Sulla scia di queste impressioni, l'odierno bevitore è tentato ad accettare un coinvolgimento ulteriore alla storiografia; più che avanzare sul terreno friabile dei *forse*, vorremmo risalire i flussi dell'immaginario in cerca di una sintonia autentica, di *profondità*. Un'intenzione di ricerca che richiede, da subito, l'impiego dell'immaginazione; entrare nelle sale degli antichi signori, respirare l'odore avvolgente del "luminoso fallimento" della loro civiltà, ma anche l'orgoglio che li porta a ritenersi superiori alle altre etnie italiche; constatare infine la *riunificante perdizione* che il vino e i suoi possibili *orientamenti* tendono a

[23] Com'era uso in Oriente: si veda ad esempio un noto rilievo del palazzo di Assurbanipal a Kuynujik, del VII secolo a.C.

suscitare.

Nell'intreccio che abbiamo descritto ed evocato grazie alla storia e non solo, è forse più facile avvicinarsi alla *Tavola degli Etruschi* poiché s'intuisce come nella loro *tensione gastronomica* (di cui il vino fa pienamente parte) si possa raccogliere una profonda istanza d'individuazione, riconoscendovi forse la nostra stessa *voce*. Alla base potrebbe esserci una pulsione non dissimile, nella sua basica vitalità, a quella che anima certe tendenze identitarie dei toscani odierni: la loro cucina spesso pungente, sanguigna e libidinosa, poco aperta alle contaminazioni...

È indubbio che *questi Etruschi* mangiassero molto e bene, diversamente dai loro vicini contemporanei abituati a pasti frugali, cotti velocemente, consumati una volta al giorno e spesso in piedi. La "misteriosa" *nobiltà etrusca,* naturalmente, ingrassava. Catullo parla dell'*obesus etruscus* e Virgilio di un *pinguis tyrrenus.*[24] Ma l'idea dell'etrusco "sparapanzato" favorisce, per compensazione, la percezione delle sue capacità di sognatore ed ideatore di soluzioni estetiche: di domatore e *dominatore dell'immaginazione*; la stessa che forse inghiottì, in una fatalistica autodistruzione, altri popoli e clan.

Del resto, negli affreschi, le figure mostrano una prestanza fisica e, per quanto possano essere affidabili i pigmenti antichi, una pelle abbronzata dal sole che poco si confà alla diceria romana del "pingue tirreno"; ma forse siamo già presi in un regime immaginario, in una specifica *sepolcralità*. Questi personaggi sembrano *innescati* dal cibo e soprattutto dal vino, preda di un'eccitazione ancestrale, un'ispirazione in cui materia e immagine si corteggiano tra loro conducendo a forme dinamiche di celebrazione: danza, canto, sesso e, per così dire, altri modi di *smaltire.* La morte? Uno stato d'ebbrezza portato alle sue estre-

---

[24] «*Ut parcus Umber aut obesus Etruscus...*», da Gaio Valerio Catullo, *Carmina*, XXXIX, 11:a.; «*Pinguis [...] Tyrrhenus.*» da Publio Marone Virgilio, *Georgiche*, II,193.

me conseguenze concettuali.

Certo feste e baldorie, nell'Aldilà ma certamente anche nell'*al di qua*, furono un'idea diffusa e motivata da occasioni di ogni genere, come testimoniano le molte pitture che tramandano dei succulenti simposi di commiato organizzati in onore del defunto. Tradizioni per noi paradossali ma in realtà largamente diffuse nella mondanità mediterranea. Uomini e donne, sdraiati sui triclini durante il banchetto, attraversavano due fasi ben distinte: quella in cui si mangiava, il *syndeipnon* greco, e quella del *symposion,* in cui si beveva e si conversava. Naturalmente si serviva vino in quantità, mescolato con molta acqua, drogato e addolcito con orzo e miele per favorire l'euforia[25] e migliorare la digestione; non di rado vi era grattugiato un pizzico di formaggio pecorino e si creava così il *kykeion*: la bevanda degli eroi.[26]

La scottiglia casentinese, da molti considerato un piatto di recondita origine etrusca.

---

[25] Kerényi pone il miele alle origini della ricerca dell'ebbrezza, distinguendo una *euforia* primordiale dall'*ubriachezza* scoperta dal consumo del vino. «Ciò che gli uomini più tardi ottennero dalla vite fu dapprima dono delle api». C. Kerényi, Dioniso, Adelphi, Milano 1992, p. 53 e seg.

[26] Enio Pecchioni, *Etruschi, il vincolo dell'unità sacrale,* Press & Archeos, Firenze 2016 (prima ed. 2012), p. 219.

Il rinvenimento nei corredi funerari del periodo orientalizzante (VIII secolo a.C.) dell'attrezzatura connessa al consumo del vino, quindi crateri, olle, idrie, brocchette, tazze con i relativi sostegni, in tombe riferibili ad esempio a eminenti donne dell'aristocrazia, suggeriscono che la gestione della bevanda fosse spesso affidata alla padrona di casa, la *mater familias* anche se, in linea di massima, questa si asteneva poi da farne uso durante il banchetto. Abbiamo quindi indizi della profonda partecipazione di una componente femminea al concerto enosofico, sotto l'ascendenza del più mediterraneo degli dei, Dioniso - protagonista negli specchi bronzei - che avrebbe avuto come sue sacerdotesse e cardini funzionali le menadi e le loro più tarde interpreti (ne parleremo entro breve). Tornano dunque alla mente alcune soluzioni ancora adottate nella presentazione dei vini etruschi: i loro nomi di donne, di schiave come di dee, da *Phylica* a *Cautha,* da *Velia* a *Turan.*

Il vino avrebbe d'altronde avuto un ruolo essenziale, in Etruria, già da epoca villanoviana, nei banchetti che rinsaldavano antichi legami tra i *clan.* Al tempo dei Villanoviani è confermato il ruolo cerimoniale e sacrale delle bevande alcoliche, sin dall'età dei *Campi d'Urne.*[27] Ecco la componente maschile, "lucumonica" o guerresca, rintracciabile in altri vini contemporanei, con le sue ataviche pretese.

La coltura della vite vinifera perfezionata dagli Etruschi già dall'VIII secolo con la produzione del vino, divenne quantitativamente così rilevante da costituire uno dei cespiti principali dell'economia della nazione. Ne è una prova che nell'inizio del IV secolo a.C. un etrusco di nome Arrunte partisse da Chiusi verso la Gallia per sollecitare e incitare i Galli a intervenire in Italia contro il reale pericolo di Roma, portando con sé carri

---

[27] B. L. Paganelli, *Le radici della religione etrusca*, Anguana, Sossano 2013, p. 241.

carichi d'olio e anfore piene del prodotto più eccellente e rinomato degli Etruschi.[28]

Le piante delle viti, condotte idealmente da Tirreno sin dall'Asia Minore, germogliano e crescono nelle *dolci colline* dell'Etruria con i loro grappoli zuccherini. Ecco dunque il vino della Tuscia scorrere a fiumi a riempire le anfore vinarie, gli orci, i fiaschi e infine le bottiglie di mezza Europa.

Cosa è cambiato, in sostanza, da allora?

---

[28] Marco Porcio Catone, *Orig. Fragmenta,* II,5 e altri.

Fufluns, l'altro Dioniso

Fufluns, definito spesso «versione etrusca» del greco Dioniso e indicato come una delle divinità più importanti dopo *Tinia/Velthune*, risulta nonostante tutto una figura complicata da delineare, sospesa tra strutturalità divina e funzionalità iniziatiche ma soprattutto scissa, con qualche sfumatura, tra un *prima* e un *dopo* la diffusione delle tradizioni e delle "mode" elleniche in Etruria. Ripercorrere la sua figura in senso mitologico e storico-artistico (certo senza pretese di esaustività) ci aiuterà a intuire la portata delle sue articolazioni concettuali, spesso rimossa o semplificata; e attiverà forse qualche nesso associativo sulla linea del parallelismo tra enosofia e mitologia (anche se per qualcuno resterà bizzarro che ogni tanto siano citati vini contemporanei come fossero materiali o documenti archeologici).

Le tracce del presunto "Dioniso etrusco" originario sono molteplici ma spesso tutt'altro che chiarificanti, mentre paiono promiscui e persino ironici i suoi accompagnamenti. Ricordato soprattutto in rapporto alla città di Populonia che deriverebbe appunto il suo nome da *Phuphluns/Fufluns*, ebbe fama in molti altri luoghi. I repertori delle iscrizioni etrusche e in particolare quelle databili al VII secolo indirizzano l'interesse verso la città di Vulci, dove il dio è appellato come *Pachies*; questa seconda definizione risulterebbe un epiteto, quasi certamente il germe del romano *Bacchus*; o sarebbe riferibile al fedele stesso, cioè all'iniziato che celebra il rito dionisiaco.[29] Essa è presente nell'enosfera contemporanea attraverso il nome di un noto vino ceretano (vedi cap. I).

---

[29] Nancy Thomson de Grummond, *Etruscan Myth, Sacred History, and Legend*, Museo di archeologia e antropologia dell'Università di Pennsylvania, Filadelfia 2006, p. 113.

Tra le più note rappresentazioni di Fufluns troviamo un bronzetto di provenienza populoniense databile al 480 a. C. circa, appartenente alla collezione della Galleria Estense ma esposto oggi a Malibù (Paul Getty Museum). Di età matura con folta barba appuntita, lunga veste e mantello ricamato, la figura avanza e tiene il braccio sinistro proteso e piegato per sostenere un oggetto, probabilmente un vasetto per bere (*kantharos*).

Molte altre sono le iconografie comprendenti Fufluns/Dioniso espresse soprattutto nella pittura vascolare e nelle incisioni sul bronzo. Su tutte *La Nascita di Dioniso* che s'incontra con una certa frequenza nell'arte etrusca, ad esempio in uno specchio del IV secolo proprietà del museo Nazionale di Napoli.[30] Un altro famoso specchio conservato a Berlino rappresenta Fufluns con Aplu e Semele,[31] mentre in altri casi il dio compare in triadi sovrapponendosi a volte, come vedremo, a Tinia/Zeus.

Dal trattato di Marziano Capella (IV-V sec.) deduciamo l'analogia tra Fufluns e il latino Libero ("successore" di Bacco/*Pachies*) che compare nella regione IX del cielo[32] (corrispondente al sudest), dedicata alle divinità della natura e altresì compatibile con la posizione espressa nel fegato di Piacenza, dove è nominato due volte come *Fuflns* o *Fuflnsl.* D'altra parte la radice dello stesso nome *Fuflu/ns* sarebbe attinente al concetto di «germoglio» (forse della vite stessa) e quindi alla ciclica rinascita della vegetazione; ma non è da escludersi un'interpretazione onomatopeica attinente ai flussi acquiferi. Il termine godrebbe di una speciale relazione con la cultura appen-

---

[30] Nancy Thomson De Grummond, *Mirrors and Manteia. Themes and Prophecy on Etruscan Prenestine Mirros*", in *Aspetti e problemi della produzione degli specchi etruschi figurati - atti dell'Incontro internazionale di studio*, Roma, 2-4 maggio 1997, Aracne, Roma 2000, pp. 68-69.

[31] *Fufluns/Paxies. Sugli aspetti del culto di Bacco in Etruria*, in *Studi Etruschi* XLVI, 1978, pp. 119-133.

[32] M. Cristofani, *Etruschi, una nuova immagine*, Giunti, Firenze 2000, p. 143.

ninica, forse attraverso la figura della divinità umbra *Puemune*, nominata nelle Tavole Eugubine, o la rarefatta figura del dio umbro Vufiune/Vofiono.[33]

La madre Semia, analogamente alla Semele greca, è talvolta ricondotta all'elemento-terra. Nel mito greco Semele muore folgorata dalla manifestazione di Zeus che aveva conosciuto solo in sembianze umane e a cui, convinta dalla sua nutrice (in realtà Hera camuffata), aveva chiesto di mostrarsi nella sua realtà. Questo fatto, strutturale al mito di Dioniso, non permette di associare *Semle* alla Grande Madre e la collega a una dimensione esistenziale prossima alla condizione umana.

Partorito prematuro, custodito e rinato dalla gamba di Zeus (quindi «nato due volte»); prelevato da Ermes e cresciuto dalle ninfe, Dioniso discese negli inferi per salvare la madre, esprimendo così l'idea universale della rigenerazione. Destinato a un'esistenza contraddittoria, non sempre accettato a pieno nell'Olimpo, subirà l'avversità di personaggi come Licurgo, Perseo, i Titani e qualche divinità, reagendo spesso con scaltre fughe e comparse altrettanto improvvise; tutto questo è stato interpretato considerando l'ostilità della cultura dominante nei confronti di un dio d'origine tracia e protagonista di segreti misteri. Ma non dimenti-

Nascita di Dioniso dalla coscia di Tinia nella riproduzione di uno specchio conservato al Museo Nazionale di Napoli (IV sec. a. C.).

[33] Aldo Neppi Modola in *Enciclopedia Italiana*, 1932, *cfr.* L. R. Taylor, *Local cults in Etruria*, Roma 1923, p. 207.

chiamo la sintonia di questa figura con i cicli della vegetazione: la decadenza e il rinnovamento, l'oltretomba e la rinascita.

La suddetta avversità fu forse meno pronunciata nelle città dei Rasna che accolsero *Dionysos* adattandolo ai paradigmi umbro-etruschi, facilitati forse dall'esistenza di leggende che legavano le origini del popolo all'Anatolia, o comunque da un certo *feeling* per l'Oriente. Almeno dalla seconda metà del VI sec. a. C. l'immagine di Fufluns appare dunque sovrapposta a quella di *Dionysos* e dotata all'incirca dei medesimi attributi. Salvo quando il nome è scritto a lato, non è possibile comprendere come fosse definito il dio da artisti e fedeli; se appunto ci si riferisse ad un "autoctono" *Fufluns* o a un più "internazionale" Dioniso, o a entrambi.

Nel corso del V-IV secolo i culti dionisiaci si affermarono in Etruria e con raffigurazioni come *Dioniso giovinetto* si diffuse maggiormente il nome greco. Ma più soggetti e iconografie si sovrapposero nelle tradizioni dei *Rasna*: Fufluns appare insieme alla madre *Semle* e alla compagna *Areatha* (Arianna), così come a *Vesuna*, a *Sethlans*/Efesto[34] e ad altre figure non sempre identificabili. In almeno due specchi *Areatha* è donata da Artemide a Fufluns in presenza di Minerva ma l'episodio non sembra avere riscontri ellenici,[35] salvo non si voglia intendere il gesto come una richiesta di sacrificio della ragazza. Secondo una versione del mito Dioniso ricevette quest'ordine poiché Arianna, sfruttata e abbandonata da Teseo a Nasso (dopo le note vicende cretesi), aveva perduto la verginità; più diffusa la versione in cui Dioniso salva e sposa la ragazza. Forse gli Etruschi avevano assunto queste ed altre figurazioni in base alla loro speciale sensibilità.

---

[34] Lucio Giunio Bruto *Ricerche di storia, religione e diritto sulle origini della Repubblica romana*, La Réclame, Roma 1988, p. 252.

[35] G. A. Mansuelli, Enciclopedia dell'Arte Antica (1965), *cfr.* Gerhard, Klugmann, Körte, *Etruskische Spiegel*, LXXXVIII e 305.

Innumerevoli sono le pitture vascolari rappresentanti Dioniso/Fufluns, le danze dionisiache e i cortei dei misteri, i gruppi di figure con donne e fauni avanzanti verso il dio (che attende spesso con la coppa alzata) in groppa a tori, brandendo serpenti, indossando maschere...ecco il seguito (*thiasos*) eterogeneo e spesso impenetrabile di Fufluns. In molti dipinti, come nel caso di alcuni crateri vetuloniensi provenienti da una serie di recente confisca (400-325 a.C.) si osserva Dioniso seduto o danzante con in mano il tirso, che è forse il suo attributo più noto dopo il calice sacro. Questo bizzarro strumento rimanda ancora una volta al simbolismo della fecondazione e, indipendentemente dalla sua diffusione nel mondo ellenico, richiamerebbe valenze italiche poiché il nome *tirso* coinciderebbe con una radice protoumbra; d'altro canto il nome greco dei *Rasna*, *Thyrsenoi*, avrebbe buona assonanza con la parola greca *thyrsos* – ma non per

Dioniso con una menade, da un Kilyx proveniente da Chiusi (IV sec. a. C.).

questo credo si possa definire gli Etruschi "utilizzatori del tirso".

Lo scettro o bastone cerimoniale compare in più forme e declinazioni; si dice fosse fatto di corniolo ma possono intrecciarvisi edera e soprattutto vite. Sulla sommità si nota quasi sempre una protuberanza più larga, generalmente interpretata come una pigna dalla quale, a dir di Euripide (ne *Le Baccanti*), scaturiva miele. A parte le interpretazioni falliche possiamo ricordare come il miele fosse spesso mescolato al vino per perfezionarne il sapore e rendere più intensa l'ebbrezza.

Tra gli attributi che compaiono nelle pitture, oltre alla patera (utilizzata anche nei sacrifici) o la cetra, sembra importante il ruolo degli *stivali/calzari*, condiviso con una schiera di figure catactonie o legate alla natura silvestre, come il cacciatore *Zagreus* (cioè *Zeusnysos*, figlio di Zeus, ipostasi dello stesso Dioniso nei culti orfici[36]), i demoni *Culsu/Culsans* (custodi delle *porte*) e *Selvans* (custode dei *confini*),[37] nominati nel fegato di Piacenza (sempre nell'ambito della sfera ctonia), ed anche il cacciatore *Atunis*/Adone. Anche *Atunis* è protagonista di misteri iniziatici: la nascita dal fusto di un albero in seguito a vicende incestuose e il legame forzato con gli inferi (nel mito, Persefone ottenne di averlo con sé per un terzo dell'anno). Il culto di *Atunis* ebbe successo tra gli Etruschi come testimoniano una nota area sacra presso Gravisca (porto di Tarquinia) e numerose iconografie (comprensive di bizzarre divinità alate). Per quanto riguarda la viticultura contemporanea, segnalerei almeno un vino laziale che porta questo nome.[38]

---

[36] Zagreo viene fatto a pezzi dai Titani inviati da Era, poiché il padre Zeus aveva ipotizzato di farlo suo successore. Zeus, oppure Semele, mangiano il cuore ancora pulsante del giovane prima di concepire Dioniso. In altre versioni sono Rea o Apollo a mettere insieme i pezzi di Dioniso per rendergli la vita.

[37] Titolare, tra l'altro, di un raffinato vino a base di Merlot prodotto da Mezzetti (Cortona), già ricordata per un *Lucumone*.

[38] L'azienda è Pacchiarotti di Grotte di Castro, che produce anche un noto Turan.

In un antichissimo mito sulle origini della viticultura, precedente forse agli stessi culti dionisiaci, Oresteo, altro cacciatore o "uomo delle montagne", vide la sua cagna partorire un ceppo di vite da cui ottenne la prima coltivazione. Il mito è forse collegabile all'apparizione in cielo di Orione (forte di Sirio, il suo *cane*) nello stesso momento in cui maturavano le uve.[39] Al contempo resta interessante la continuità del riferimento a figure legate alla caccia, quindi alla natura silvestre ma anche alla selvaggina, da sempre associata al consumo di vino - a sua volta connesso ad un'idea di sacrificio che ha attraversato secoli e religioni.

Sorvolando sulle molte figure attinenti ai miti mediterranei dedicati all'origine della vite - per le quali servirebbe un'intera trattazione - è da notare come, nell'immaginario etrusco, i personaggi collegati al vino corrispondano spesso a dei e demoni psicagogici che conducono l'anima, con il loro passo agile ma sacrale, nel confronto con il post-morte e che compaiono in specchi e raffigurazioni funerarie. A tal proposito è possibile che il mito di Dioniso salvatore di Semele, come quello di Adone e di Afrodite si accordassero con un mito etrusco o di recondita origine appenninica, caratterizzato da divinità ctonie di matrice umbra-italica.[40]

Fufluns/Dioniso sarebbe dunque relazionabile a quello "strato intermedio" tra suolo e inferi in cui operano incessantemente, coadiuvati dalle divinità della terra fertile (Demetra/Proserpina, Uni/Turan, ecc.), i vettori, generalmente maschili, della fecondazione e della rinascita. Con il suo peculiare dinamismo il dio del vino tende da un lato al rigoglioso mondo dei vivi e dall'altro all'Ade, risultando protettivo per l'adepto nell'ambito della dipartita terrena quanto nelle ritualità più estreme; e così in quelle

---

[39] Sulle varianti del rapporto tra Orione e il tema vinifero/dionisiaco, si veda anche C. Kerényi, *cit.*, p. 59 e seg.

[40] B. L. Paganelli, *Le radici della religione etrusca*, Anguana, Sossano 2013, p. 248 e seg.

pratiche magiche che con lo scopo d'intervenire sul reale solleticano i delicati equilibri dell'Aldilà. Attinente ad una sfera pienamente iniziatica sembra essere l'attributo della nebride, la pelle di cerbiatto (sacrificato) indossata da Dioniso o anche da *Vesuna*, sua ambigua compagna in alcune rappresentazioni italiche (la divinità, d'origine umbra, è dotata di corona con corna/luna crescente). L'immagine della figura coperta dalla pelle è al contempo familiare al mondo celtico: soprattutto attraverso la figura iniziatica di Cernumno.

In molti casi constatiamo dunque la vicinanza di Fufluns ad un sottoinsieme di figure mitologiche che si ritengono parte del suo seguito, distinguendosi però dal "pubblico", essenzialmente decorativo ma non privo di caratterizzazioni, delle generiche menadi e satiri. Tra questi ultimi potremmo notare il vecchio Sileno, ebbro e profetico, forse in groppa ad un asino come capita di ritrovarsi al dio Efesto (quando viene ricondotto da Dioniso sull'Olimpo, talvolta con la partecipazione di *Maris/Laran*[41]). O potremmo intravedere il figlio stesso di Dioniso e Arianna, *Oinopion* (ne parla una leggenda originaria di Chio), che da giovane coppiere del padre si trasforma (e deteriora) in sileno barbuto e claudicante[42]... come dire che l'ebbrezza non "risparmia" nemmeno il figlio del dio.

Ma è nelle iconografie prettamente italiche che vediamo comparire quei demoni capaci di veicolare significati tanto remoti quanto incisivi all'intuito. Oltre ad *Areatha* o *Vesuna*, la compagna di Bacco sembra essere *Catha/Cautha*, che compare specialmente in età ellenistica. Venerata a Tarquinia e a Pyrgi ed attinente al potere dell'*alba*, sorella del sole o astro essa stessa, è forse collegabile a *Phersipnai* con conseguente sovrapposizione

---

[41] G. A. Mansuelli, *cit.*

[42] Si assiste a questa mutazione nelle iconografie delle pitture su ceramica nel corso del V-VI sec. a. C. (A. Gallina in *Enciclopedia dell'arte antica*, Roma 1963).

di Ade e Dioniso e relazione, ancora una volta, con il mondo sotterraneo.[43] Da ricordare, per quanto ci riguarda, l'esistenza di un vino che porta proprio il nome *Cautha* (vedi cap. I), non a caso simboleggiata da un sole fiammeggiante (in etichetta).

Su un cratere proveniente da Pyrgi ritroviamo inoltre una figura indicata come *Fuflunusura*, traducibile come "appartenente alla cerchia di Fufluns" in relazione quindi a *Sura/Suri*, venerata insieme a *Cavatha* in una specifica area sacra nella parte più a sud del santuario di Pyrgi.[44] Mentre per Cavatha si propone ancora l'analogia con Persefone, Suri sarebbe identificabile con Apollo delfico (la città etrusca di Cere ebbe un suo tesoro a Delfi) con ennesima conferma di un rapporto tra *Aplu* e *Fufluns* passante dal concetto-divinità *Suri,* così come il dono profetico può manifestarsi in entrambi i fratelli. Per Fufluns questa capacità sarebbe confermata anche dalla presenza della patera, condivisa almeno in un caso con *Vesuna* che, nell'immagine di uno specchio, regge il contenitore sul fondo del quale venivano letti gli eventi futuri.[45]

In alcune pitture compaiono assieme a Dioniso anche una o più pantere: uno degli animali preferiti dall'eccentrica divinità il cui carro volante appare a volte trainato da questi felini.

Insomma, nel seguito di Fufluns, autentica "star" del substrato antico-arcaico, sembra rivivere, oltre certi eccessi quasi caricaturali, un buon numero di figure polifunzionali, o un piccolo *pantheon* interno: una serie di "ministri-ombra" degli Dei olimpici. Certo è che in questa figura e nella banda delle sue nume-

---

[43] Secondo un inno orfico, quando Dioniso è assente, si ritiene che si trovi presso Persefone. Si veda Mircea Eliade, *Storia delle credenze e delle idee religiose,* Sansoni, Firenze 1979, vol. 1, p. 388 e seg.

[44] Andrea Ciacci, Paola Rendini, Andrea Zifferero, *Archeologia della vite e del vino in Toscana e nel Lazio. Dalle tecniche dell'indagine archeologica alle prospettive* della *biologia molecolare*, All'Insegna del Giglio, Firenze 2012, p. 287.

[45] B. L. Paganelli, *cit.*, p. 246-247.

rose compagne e compagni potrebbero mostrarsi le tracce di una cultualità primigenia, lentamente accantonata a favore di iconografie più limpide e rassicuranti (come in una scena di colloquio con *Hercle*, *Svutaf*-Eros e *Vesuna*, o altre in cui il dio appare efebico e dotato di cetra[46]). E risulta significativo che proprio il *dio del vino* sia stato a capo di questa antica tribù regnante.

Se sui vasi e sugli specchi si manifestano personaggi e attributi con una certa libertà, nella scultura e in particolare nella plastica degli idoli/*ex voto* si osserva una connaturata sintesi concettuale che potrebbe suggerire le potenzialità più ancestrali e irriducibili di Fufluns-Dioniso. Tuttavia anche in questi esempi non mancano ambiguità e sovrapposizioni.

Numerose statuette etrusche mostrano l'immagine di Tinia in età matura, con lunga barba e capelli spesso raccolti sulla schiena, in procinto di avanzare (secondo il tipico modello tardo-arcaico) e offrire presumibilmente il *kantharos* (se non la saetta, attributo tradizionale del dio). La posa è del tutto analoga a quella del già citato bronzetto populoniense ed è confermata in altre raffigurazioni rapportabili a Fufluns, come un idolo votivo di età tarda (del III sec. a. C., conservato alla Biblioteca Nazionale di Parigi[47]). Si veda ancora una statuetta del 480 ca. conservata a Los Angeles (rappresentante Fufluns), rapportandola ad un idolo dello stesso periodo (in cui si riconosce Tinia) proveniente forse dal deposito votivo del Falterona, il Lago degli Idoli e conservata oggi a Modena.[48] Sembrano entrambi uomini giovani, associabili al *concept* di entrambe le divinità. Queste evidenze ci introducono all'annosa questione della sovrapposizione tra il "Dioniso" e il "Giove" etruschi.

---

[46] G. A. Mansuelli, *cit.*, *cfr. Etruskische Spiegel*, *cit.*, v, 35.

[47] Mario Torelli, *Storia degli Etruschi*, Editori Laterza, 1981, fig. 127, p. 268.

[48] Mario Torelli, *Gli Etruschi*, Bompiani, Modena 2000, p. 604.

Tinia o Fufluns?
L'ambiguità nell'ebbrezza

La componente dionisiaca di Tinia risponde ad una relazione di reciprocità tra queste figure poiché il maaggiore dio etrusco, diversamente dallo Zeus greco, si mostrava anche in forme giovanili e questo fu senz'altro il caso di alcune importanti icone; come probabilmente di riproduzioni legate a forme di venerazione frugali, proprie della dimensione quotidiana e per questo assai diffuse nella popolazione etrusca.[49]

Già in epoca remota, idoletti rappresentanti un *fulgurator* (scagliante il fulmine) non presentano l' "attributo" della barba e vista la cura di altri dettagli è difficile pensare che si tratti di semplici approssimazioni. I *Rasna* desideravano spesso mostrare un'immagine giovanile, tutt'altro che pedante del loro dio e del loro popolo: pensiamo anche alle vivide pitture sepolcrali. È emblematico il caso di uno specchio d'argento proveniente da Bomarzo, datato alla metà del IV secolo, recante l'immagine di *Tinia* con *Aplu* e *Turms* ai lati. Il padre degli Dei appare imberbe, giovanile e sorridente, tanto che potremmo considerarla una buona rappresentazione di Fufluns stesso (anche Dioniso ha spesso una collana fatta di bulle o denti) se non fosse per l'iscrizione che chiarisce il nome del dio e per l'attributo del fulmine nella mano destra (ma forse, tirso e fulmine possono prendere forme alquanto somiglianti[50]).

---

[49] A tal proposito Plinio, oltre a testimoniarci della statua di cui ci occuperemo tra qualche pagina, racconta di tazze intagliate nel legno di vite (una vista a Marsiglia) e persino di scale i cui scalini sono fatti dello stesso materiale (a Cipro). Plinio, *Naturalis historia*, XIV, II,9.

[50] Penso ad alcune statuette come quella del Peglio (Cortona, *MAEC*, VI-V sec. a.C.) in cui la sintesi delle fiamme del fulmine può ricordare le brattee di una pigna.

Due rappresentazioni bronzee di Fufluns. A sinistra, imberbe (Museo Nazionale di Parigi, III sec. a.C.). Sopra, la nota statuetta populoniense (V sec. a.C.) esposta a Malibù.

L'immagine è coetanea di altre che mostrano barbuti sia il capo degli dei, sia Fufluns/Dioniso. In un noto tipo di Zeus diffuso a Orvieto, a Faleri e lungo la costa ceretana, Tinia appare giovanile ma con barba (non lunga): nel corso del IV secolo le "acconciature" variano e s'intuisce l'esistenza di mode locali,[51] oltreché una ricerca di eleganza e attualità (forse un sintomo del corso della decadenza dell'identità nazionale-etrusca); in altri *Dreifigurengruppe* è accostato a una menade e un satiro (un "*thiasos* abbreviato") mantenendo la sua barba appuntita, secon-

[51] Mauro Cristofani, *Civiltà degli Etruschi,* Electa, Firenze 1985, p. 284.

do uno schema arcaico d'ispirazione attica.[52]

È inoltre significativa la scarsità di composizioni in cui Tinia e Fufluns compaiono insieme, come se i due fossero quasi interscambiabili. Del resto, il *Dionysos* etrusco partecipa a triadi dominate, almeno in certi casi, da misteriose intenzionalità. Si pensi agli specchi in cui appare con Castore (*Kastur*) o Giasone (*Eiasun*)[53] o nei frangenti in cui, insieme a *Menvre*, incontra *Artunes*[54]; a volte, intendere il protagonista come l'uno o l'altro *numen*, può portare ad interpretazioni diversamente calzanti.

È possibile che l'attributo della barba non fosse così carico di significati e che definisse diversi momenti (uno più acerbo, l'altro più maturo) di un percorso che riguardava l'iniziato ai misteri come l'essere umano nella sua parabola vitale. Ma possiamo forse dedurre qualcosa di più dalla sovrapposizione delle due entità, Tinia e Dioniso, il padre e il figlio; fosse solo una tendenza emergente qua e là, secondo un'inconscia ricerca di un senso identitario; come dire che il capo delle divinità degli Etruschi, già noti nel mondo antico per il loro vino, non poteva non essere dionisiaco per eccellenza. Su questa linea, certo un po' sognante, il *khantaros* può sostituirsi al *fulgor*; ma a pensarci i due concetti non sono in contraddizione tra loro, almeno nel senso dell'accensione di un fuoco (dal greco πῦρ), un processo di resurrezione energetica - se non propriamente iniziatico.

In senso generale, si può dire che la capacità di fusione tra le figurazioni divine è una costante della cultura etrusca e difatti il dio più famoso e venerato, quel *Velthune/Voltumna* patrono della Dodecapoli, è per definizione mutevole, ambivalente, legato ai cambiamenti delle stagioni e delle evenienze, dunque lui stesso

---

[52] Laura Ambrosini, *Specchi dispersi da una tomba di Corchiano* in *Archeologia classica, rivista dell'Istituto di archeologia della Università di Roma*, vol. 48, Roma 1997, p. 255.

[53] G. A. Mansuelli, *cit.*

[54] Nancy Thomson De Grummond, *A Guide to Etruscan mirrors*, Archaeological News, 1982, p. 112.

"praticamente dionisiaco". Al contempo, per la sua notevole diffusione, *Velth* si sovrappone a Tinia e se ne parla spesso come di un epiteto di quest'ultimo. Nel mito più noto (ma forse tardo) in cui si trasforma in donna per avvicinare *Pomona*, pare chiaro il suo legame simbolico con la vegetazione, mentre il suo stesso nome sarebbe legato al latino *vertere* quanto al sanscrito *vàrtate*: ancora natura, ancora improvvise epifanie, ancora Oriente, in un contesto che pare improntato, appunto, su una sintesi dinamica.

Ma quale poteva essere dunque, in questa mescolanza di nomi e immagini - per così dire, in questo *blend* - la radice iconografica del Fufluns originario? In realtà non sappiamo nemmeno se il nome etrusco/italico precedesse in qualche modo quello greco, e se non si debba rimandare entrambi a un'ulteriore figura-divinità primordiale di matrice vicino-orientale. Pare insomma che, per comprendere la realtà del singolo "podere" dove il dio è oggetto di venerazione, si debba comunque allargare il fuoco al Mediterraneo originario.

Concentrandoci sui prototipi e sulle modalità attraverso cui la divinità o le sue ipostasi erano chiamate a presenziare alle attività della vita sociale e quotidiana (dunque non attraverso la codificazione tradizionale dell'*ex-voto* con la sua economia simbolica peculiare), possiamo immaginare che il dio *Fufluns,* e forse qualsiasi figura investita dalla più ancestrale componente dionisiaca, "gradissero" esser raffigurate su *legno di vite* o di altro albero da frutto; se non proprio scolpite in questi legni speciali, antichi, capricciosi.

La scultura in legno ricopre in antichità una funzione speciale quanto essenziale. Già a Nasso, dove Dioniso era venerato da tempi antichissimi, esistevano maschere lignee drammatiche che rappresentavano il dio furente definito con l'epiteto di *Bacheús* (in legno di vite) o *Meilíchios* (in legno di fico).[55] E proprio a Nasso c'è chi pone le origini dello stesso nome *Fufluns*.[56] Le icone in legno, dette spesso *xoana*, coinvolgevano generalmente divinità legate ai cicli della natura, da Era ad Ercole, da Afrodite a Cibele e la tradizione greca le fece risalire all'arte primigenia di Dedalo. Deperibile ma organica, immagine ricavata dal legno si caricava di un fascino primordiale; forse ancor più se quel legno aveva donato l'olio per le unzioni sacrificali, o l'uva per il miracolo della vinificazione. Nel caso di certi *xoanon* si narra di tronchi ormai bisunti adorati da chissà quante generazioni: idoli talmente usurati da poter trasmettere all'istante l'idea fondamentale del divino-terrestre o, nell'attività iniziatica, l'ispirazione basica dell'estasi e della fecondazione.

Legni di vite utili alla realizzazione di vere e proprie statue erano rari da ottenersi nelle misure o quantità necessarie, sebbene nel mondo antico la vite somigliasse assai più a un alberello[57] e non fosse necessariamente organizzata in filari simili a quelli odierni. La pianta con il suo legno sfilacciato, curvoso, rampicante o cespuglioso cresceva dritta, con supporti e tutori, e poteva ricordare le immagini vegetali dei racconti mitologici, da cui scaturivano semi e frutti "magici" identificati spesso con ninfe e altre figure ammallianti.

Per crescere più alte e più forti le viti erano piantate quasi sempre vicine o appoggiate ad altri alberelli, specialmente piante da frutta, che al contempo distoglievano gli insetti e gli uccelli dall'uva. Le piante venivano a loro volta circondate da fitte siepi, forse di alloro o d'olmo, per essere protette dagli animali alla ricerca di pascolo. Fufluns e la sua vite erano dunque accompa-

---

[55] Walter Burkert, *La religione greca*, Jaka Book, Milano 2010, p. 326.

[56] L'associazione, secondo altri un po' ardita, passa dal nome del vino *βύ(ι)βλινος*. Aldo Neppi Modona, *cit.*

[57] Werner Keller, *La civiltà Etrusca*, Garzanti, Milano 1971, p. 59.

gnati da altre entità, i *tutori vivi* di cui parla l'archeologia della viticultura, gli alberi dei frutti del sole (*Cautha*, *Vesuna*, *Areatha*) e le piante urticanti analoghe a menadi e satiri, atti a custodire il culto e ricavarne reciproco vantaggio.[58] Qualcosa di analogo a ciò che vediamo in certi specchi: Dioniso, barcollante, retto in piedi da giovani dee e sileni...

Possiamo immaginare che alla morte dell'alberello (dopo circa mezzo secolo o più di produzione, quindi in probabile sincronia con la generazione umana corrente) fosse usanza conferirgli l'immagine del dio e renderlo, diciamo, una sorta di *xoanon* (seppure in questo caso il concetto non sia usato in modo assolutamente proprio). Questi piccoli culti, se esistettero, si consumarono a livello locale, in piccoli sacelli o in ambito poderale-familiare. E alcuni vignaiuoli odierni non avrebbero esitazioni a sottolineare l'importanza di certe conservazioni (se non scaramanzie), o d'aver sentito parlare di usanze del genere da qualche anziano.

Un esempio di legno di vite centenaria.

Proprio nell'antica Etruria esisteva almeno un caso di famosa statua lignea. A Populonia, la città di *Fufluns*, si venerava un simulacro intagliato nel legno di un'antica vite (o rivestito di esso); è Plinio il Vecchio a testimoniare questa

[58] Vengono in mente le rose poste a inizio filare dai vecchi vignaioli dello scorso secolo, per dedurre la presenza di malattie o particolari insetti.

presenza nella *Naturalis historia*.[59] La statua, «*Jovis simulacrum*», era forse ricavata da un unico legno e rappresentava presumibilmente Tinia o Velthune. È possibile che il Tinia di Populonia avesse sembianze compiute e riconoscibili, o che fosse invece "consumato" insieme al suo deperibile legno. Nella misura della sua inessenzialità, avrebbe potuto presentare qualche tipo di coincidenza con quel Fufluns da cui la città prendeva probabilmente il nome. La stessa testimonianza di Plinio, d'epoca tarda, potrebbe suggerire e al contempo travisare antiche funzionalità, riportando la tradizione populoniense al "solito" Giove. Difficile pensare che il dio adorato nella città di *Phuphluna* non fosse investito dalla componente dionisiaca che abbiamo già osservato.

Probabilmente Tinia/Velthune, erede diretto dell'antichissima divinità della rigenerazione, ebbe nei suoi culti una componente "dionisiaca" già molto prima che *Dionysos* sbarcasse in Etruria con i suoi *misteri*. In onore di Tinia, come di Fufluns, venivano già eseguiti sacrifici taurini e lo stesso Toro è considerato animale consacrato a Dioniso. Il Toro è animale iniziatico per eccellenza già in Asia Minore e nell'Oriente mediterraneo, ponendosi agli albori della civiltà.

Di fatto questo antico nume trovò nelle tradizioni, già abbozzate, della cultura della *vitis silvestre* e del suo *temetum*, gli strumenti metaforici idonei alla manifestazione di idee universalmente valide: la rinascita attraverso la macerazione e la fermentazione; la salvezza stessa dell'anima del devoto e del *bevitore* in senso lato. D'altronde nel ripercorrere la complessità della figura mitologica e delle sue relazioni, e nel constatare la sua parziale quanto vitale *inevasità*, abbiamo avuto l'impressione d'occuparci del più complesso e sublime dei *vini*: con la quantità

[59] Plinio, *Naturalis historia*, XIV, II,9.

dei suoi uvaggi, l'ambiguità della struttura, l'articolato registro di sentori e sapori; e con il rischio, a tratti, di perdere in oggettività – un momento forse necessario che avranno vissuto, almeno una volta e un tempo, anche i più attenti degustatori nel rapportarsi a vini "mitici".

L'evocazione di Fufluns/Dioniso può collegare una visione notevolmente ampia e fondamentale all'universo locale e cronologico, sino al microcosmo del singolo podere storico. A tal proposito è bello citare una filastrocca appenninica, già inoltrata da Leland e Camporeale, che richiama ancora, non sappiamo con quale effettiva fondatezza, proprio Fufluns o *Faflon*:

Faflon, Faflon, Faflon!
A voi mi raccomando!
Che l'uva nella mia vigna
è multa scarsa,
a voi mi raccomando,
che mi fate avere
buona vendemmia!

Faflon, Faflon, Faflon!
A voi mi raccomando!
Che il vino nella mia cantina
me lo fate venire fondante,
e molto buono,
Faflon, Faflon, Faflon![60]

Ma proprio nelle iconografie arcaiche, che si tratti di Fufluns "il maturo", di Tinia "il giovane" o di un "generalistico" Dioniso, troviamo forse l'immagine più semplice e potente: l'avanzare alzando il *kantharos* compiendo un'offerta e/o un "cosmico"

[60] G. Camporeale, *Gli Etruschi Storia e Civiltà*, Utet, Torino 2005, p. 209.

prosit. Sorvolando sugli antichi significati religiosi o iniziatici, si tratta di un'immagine dal valore assiale, capace d'*orientare* gli spiriti della collettività.

L'io si rivolge all'altro capo di un diagramma ideale che antepone il paesaggio all'individuo; la Storia è centrata nel suo cuore mitico dal vettore dell'enosofia e trasuda sangue ai lati nella tradizione, nella toponomastica, nella psichizzazione della Natura: roccia, flora, acque, fauna...la cui essenza è posta ora innanzi al bevitore-protagonista.

Il gesto magico-artistico dell'offerta, l'alzare il calice al tramonto innanzi al paesaggio soleggiato - e il vino che s'agita e ravviva - proiettano l'anima nel cosmo, rendono l'uomo al territorio e al suo *brulicare* (e viceversa) evocando l'esperienza definitiva della lucidità dell'ebbrezza.

Ma quanto, oggi, riusciamo a rendere realistica e percorribile questa *resa,* che nell'antichità si attualizzava anche in drammatiche ritualità?

L'OMBRA DELLE RITUALITÀ

Per quanto i vini potessero essere diversi e definibili, per quanto fossero mescolati con varie essenze e spezie, il bevitore antico non era probabilmente responsabilizzato, com'è invece l'individuo contemporaneo, nel senso del *gusto*, quindi del «riconoscere le differenze» analizzando le caratteristiche delle singole produzioni vinicole.

Del resto avere a che fare con il vino non significava solo berlo e gustarlo e non ci si limitava a godere di quell'attimo d'euforia o di riscatto esistenziale che la tensione alcolica può produrre; ma significava spesso confrontarsi con un insieme di convenzioni tradizionali e cultuali, sovrastrutture dall'origine insondabile che precedevano e seguivano la semplice bevuta e che implicavano il dominio della propria immaginazione. Si passava inoltre, pure in modi molto diversi rispetto ad oggi, da un entusiasmo contenuto ad un pragmatismo rituale. Nonostante questo, ad una più attenta analisi, i *nuclei* di ebbrezze così diversamente gestite potrebbero risultarci compatibili; e gli stessi riti potrebbero rivelare la loro impronta essenziale nel superamento della cronologia della singola epoca. Per questi motivi ci occuperemo ancora di antichità, in cerca d'analogie con il presente e con le sue *visioni*.

In antichità, il vino lo si mescolava con *di tutto un po'* pur d'appropriarsene simbolicamente; si versava e consumava secondo gesti antichi e tipizzati; si offriva in parte a qualche divinità e si brindava alla sua fortuna; si usava e sprecava per giocare al cottabo; si gettava sul fuoco per spegnere i roghi funebri;[61] si

[61] Se ne legge nel testo della *Mummia di Zagabria*, Benda IX, striscia 29: «*nacumaisma hinthu vinum trau prucuria*» (dopo la liturgia il vino versato dalla grande brocca...). L'usanza è

adoperava in ogni genere di cerimonia sacrale fino ai rituali iniziatici che, sfruttando anzitutto la sospensione delle inibizioni prodotta dall'ebbrezza, introducevano l'io ad una sfera d'azione ulteriore alla quotidianità, rielaborando una funzione metaforica del territorio, distorcendone la percezione secondo speciali fattori estatici - tanto che il *monte* o *bosco* stessi divenivano "divinità" e "templi".

Da un lato l'inevitabile varietà di soluzioni adottate nell'arco di molti secoli, dall'altro la scarsità di testimonianze su eventi pressoché segreti (e in seguito "maledetti"), rendono complesso chiarire i tratti essenziali della maggior parte di queste tradizioni. Certamente l'utilizzo del vino avveniva su vari livelli prendendo più o meno importanza, determinando una struttura d'analogismi che avvolgeva il contesto in senso onirico-immaginario quanto biologico-sociale.

C'erano ovviamente le feste dionisiache che coinvolgevano il popolo secondo cadenze definibili, almeno per quanto riguarda il mondo greco-ateniese. In Grecia, dal tardo autunno alla primavera inoltrata le *Dionisie campestri*, le *Lenee*, le *Antesterie* e le *Grandi dionisie* di cui ci parlano gli storici erano caratterizzate da cortei danzanti, falloforie, fuochi e rappresentazioni evocanti la fugace apparizione o la tetra scomparsa del dio.[62] Si trattava anzitutto di occasioni per manifestare ciò che era generalmente trattenuto, relegato dalle convenzioni sociali in *abissi* privati e personali. Si assisteva a competizioni di bevitori, giochi d'equilibrio su otri, gare di cottabo, tutto caratterizzato da un'euforia che ricorda la vitalità delle pitture sepolcrali etrusche, sebbene delle *dionisiache* dei *Rasna* non si sappia nulla o quasi.

---

ricondotta alle tradizioni greche assunte dagli Etruschi, in riferimento speciale al rogo funerario di Patroclo (Iliade).

[62] Otto definisce Dioniso «colui che arriva» e Kerényi teorizza sulle «forme dell'arrivo». C. Kerényi, *cit.*, pp. 142-143.

L'evocazione di un rapporto con l'aldilà, uno dei fattori essenziali dei misteri, risaltava già nelle cerimonie pubbliche. L'ultimo giorno delle Antesterie era dedicato ai morti e lo stesso Dioniso veniva a suo modo ritenuto "morto"; a Delfi esisteva una tomba di Dioniso e ad Argo, in un rituale di cui ci parla Plutarco,[63] il dio "tornava" dal mondo dei morti. Inoltre la sua simpatia per Proserpina è confermata, come abbiamo visto, da alcune iconografie.

Eventi miracolosi testimoniati nell'antichità come sorgenti che fanno sgorgare vino, anfore lasciate vuote e trovate piene durante la notte, alberi e vigne che producono frutti in un sol giorno, ricordano miracoli e relativi sincretismi di cui sono protagonisti, ancora nel primo Cristianesimo, personaggi carismatici che raccolsero istanze e cultualità arcaiche. Per quanto concerne la Toscana/Etruria, penso ad una schiera di santi e beati taumaturgici, alle loro sorgenti, ai miracoli derivati dal contatto o dal passaggio delle reliquie. D'altra parte, guardando al *mistero* cristiano e alle sue analogie con la tematica dionisiaca, non possiamo non ricordare il significato salvifico e rivelatorio del culto del bambino, di una madre umana quanto divina (*Semle*) e l'idea di una resurrezione dopo le violenze (penso a *Zagreus* fatto a pezzi dai Titani). Non si cerchino soluzioni sorprendenti a questo analogismo, ma si guardi a un fatto comune ad altri "salvatori": l'evidenza di una vicinanza di Dioniso alla condizione umana e alla sua *finitezza*, con l'apparato religioso-sacrale che coglie i cicli della natura in senso lato, evocando una qualche rigenerazione.

Questa familiare quanto misteriosa potenzialità che il popolo percepiva con stupore, con partecipazione ma anche con uno speciale timore, era dominata attraverso i riti e certamente, nei

[63] Plutarco, *De Iside,* 35.

misteri, s'intendeva veicolarla, probabilmente in qualche senso escatologico. Qui il vino, il suo utilizzo e consumo, prendeva chiaramente un ruolo essenziale.

Esisteva anzitutto un impiego cerimoniale che possiamo ancor oggi facilmente comprendere o almeno intuire. Nel caso dei sacrifici, ad esempio, si consacrava spesso l'animale versando sulla sua testa del vino.[64] Nell'iniziazione era utilizzato per "battezzare" il novello adepto. Si beveva inoltre fino all'eccesso allucinatorio.

Ecco dunque il lavoro dei gruppi iniziatici, con le loro attività interne e i loro riti privati che animarono, certamente, molte "compagnie" e *clan* anche in Etruria. Il vino pare implicato come "acceleratore" per rappresentare l'attivazione delle potenzialità collettive; questo ruolo prendeva lentamente campo partendo dal banchetto e dal *syndeipnon*. Già la *craterizzazione* ed altre usanze preparatorie, facenti da contorno al *simposio*, inscenavano una lenta accumulazione energetica. Il simposiarca eletto aveva il compito di miscelare vino e acqua (appunto nel *cratere*), attribuire le coppe, i posti, nonché, nei casi più sofisticati, scegliere i soggetti tratti dalla tradizione su cui ciascuno a turno doveva esprimersi, o recitare poesie, o cantare e suonare.

Durante il simposio, attraverso il dialogo, si rafforzavano i vincoli tra gli uomini e tra le diverse famiglie: un momento sacro e inizialmente distensivo, senza grandi contraddizioni con una *cena* o un *drinking-party* contemporanei, ma reso certo più complesso dalla presenza, in filigrana, del dio/demone e di tutto ciò che ne consegue. La comunanza *nel sacro* non era semplice empatia o *feeling*, ma poteva anticipare la misteriosa connessione

---

[64] Nel caso di alcuni rituali etruschi si trattava forse di un piccolo maiale o di un cane, a seconda delle divinità evocate (si veda anche, per una ricostruzione, *Aisna, il rituale etrusco secondo il Liber Linteus*, in religionidelletruriaantica.wordpress.com). Nei culti dionisiaci, qualora la cattura dell'animale non avvenisse durante l'*oreibasia*, si trattava probabilmente di un capro o di un animale selvatico.

con il dio; così, nel caso dell'iniziato, a bere era già il dio, a mangiare era il dio...e lentamente ci si allineava nella direzione di un incontro/possessione effettivi. Un fatto generalmente incomprensibile ai bevitori contemporanei, salvo grossolane semplificazioni o casi molto speciali, o estremi (che non ci interessa d'indagare ora).

Oltre all'utilizzo come "aggregante" nelle relazioni personali esistevano dunque circostanze assai più crude e avvolgenti in cui il vino trovava implicazioni. Molti simposi erano probabilmente introduttivi a una ritualità di gruppo, a "rappresentazioni collettive" dotate di organica vitalità. Forse di queste possibilità d'evocazione resta ancor oggi un'ombra - o uno spettro - che si aggira "tra le chiacchiere" dei nostri banchetti, talvolta nella selva delle speculazioni esistenziali o propriamente enosofiche, percepibile a molti e mai dichiarato se non attraverso la fantasia e quindi il sogno, l'arte o la letteratura. Ma restiamo ancora un po' sul passato.

A proposito dei misteri dionisiaci esistono alcune testimonianze antiche, soprattutto attinenti all'iniziazione dell'adepto: quelle di Demostene e di Erodoto;[65] vi è inoltre la tragedia di Euripide, *Le Baccanti*.

Sembra che il rito d'iniziazione prendesse avvio con la lettura degli *hieroì logoi*, testi sacri cui fa cenno Demostene nel noto passo di una delle sue orazioni, in riferimento ai culti di Sabazio/Dioniso ad Atene. Nel racconto compaiono da subito indicazioni formali e direi funzionali, come l'utilizzo di alcuni verbi non immediatamente comprensibili.[66] *Nebrisare* è uno di questi. Della nebride abbiamo già parlato ma è difficile chiarire come fosse utilizzata, se si trattasse di una semplice vestizione

---

[65] Demostene, *Sulla corona*, 259; Erodoto, *Storie*, IV,78-80.

[66] Per una sintesi in tal senso indico anche l'ipertesto *Il Rito dionisiaco*, *http://ospitiweb.indire.it/~copc0001/dioniso/rito.htm*

dell'adepto o di un momento più crudo e complesso, consistente nel sacrificio di un animale, nel recupero della sua pelliccia e forse persino nel consumo della carne cruda. Questa forma di *omofagia* si rifletterebbe nel mito secondo cui Zagreo/Dioniso, prima di sconfiggere la morte, viene divorato dai Titani.

Il momento della purificazione (*katharmos*) resta aperto a diverse interpretazioni; sulla base di altre esperienze iniziatiche antiche potrebbe trattarsi del cospargimento del corpo dell'iniziato con farina e/o argilla, che lo faceva ricomparire d'un colore biancastro: un "azzeramento" dei caratteri per favorire la conseguente rinascita.

Attraverso ulteriori formalità si giungeva alla danza iniziatica, in cui le menadi potevano essere protagoniste, circondando l'iniziato o anticipando il suo passaggio nel corso di una processione, come si può immaginare leggendo Erodoto. Quest'ultimo racconta l'esperienza di un re scita che, iniziato e coinvolto nei riti di Dioniso *Baccheios*, pareva preso dalla follia e baccante lui stesso.

Le menadi/baccanti danno vita ad un *thiasos primordiale*: il seguito orgiastico il cui *incipit* viene posto per ovvie ragioni ad Est. L'impressione di un legame con l'Asia Minore era evidente soprattutto per l'uso di caratteristici tamburi e strumenti che accompagnavano le danze. Per quei *Rasna* che già avevano mostrato simpatie per l'Oriente dovettero risultare familiari i culti dionisiaci e ben comprensibile l'analogia applicata tra le divinità occidentali e quelle vicino-orientali (operazione concettuale ben chiara agli antichi, come si deduce da Erodoto). Nel mito, lo stesso Dioniso aveva viaggiato in India insegnando la coltivazione del vino alle popolazioni locali e questa sua impresa era celebrata attraverso l'epiteto di *Sabazio* - in realtà un personaggio mitologico d'origine Frigia, anche lui fatto a pezzi dai Titani, finendo per sovrapporsi alla figura di Zagreo/Dioniso -.

Le baccanti sono rappresentate spesso con il capo inclinato all'indietro e così le descrive Euripide. Nella danza si dimenano e scuotono la testa in circostanze che sono state definite *isteriche*. Sulla prospettiva iniziatica, si parla del raggiungimento di uno stato di *trance* accompagnato da allucinazioni. Secondo la tragedia di Euripide questa fase prende avvio con una corsa nei boschi montani, nello specifico sul monte Citerone ma sembra che l'intero Parnaso fosse considerato teatro dell'evento. Forse ogni città aveva luoghi adibiti a simili scopi rituali ma il Parnaso era senz'altro il più importante.

L'usanza della *corsa* era definita *oreibasìa* e si svolgeva in inverno, quindi all'inizio della rinascita del sole. Vi prendevano parte *dionisiaci* provenienti da tutta la Grecia che ascendevano in lunghe processioni, vestiti di bianco senza nessuna distinzione sociale, alla luce di torce imbevute dello stesso miele versato nella craterizzazione, di alcool e ginepro.[67] L'incedere era ritmato da rumori percussivi e di cembali che nel riverbero montano assumevano la gravosità del tuono e del divino, sino a far "tremare" il terreno sotto gli stretti sentieri.

Sembra che le donne già dalla prima notte si dedicassero alla danza estatica nella ricerca disperata della possessione del dio. In certi casi le iniziate anticipavano probabilmente le processioni recandosi in stazioni a diversa altitudine, creando dei contesti dedicati a funzioni e necessità specifiche, forse relative a singoli gruppi iniziatici e alle loro "richieste". Qui si consumavano ritualità segrete, forse vaticini, altri sacrifici, accoppiamenti. Molti erano preda di allucinazioni. Durante le cerimonie, secondo le credenze, i tirsi spillavano miele e il vino sgorgava dalla terra creando ruscelli. Con il trascorrere della notte le pregnanti atmosfere, i suoni, gli odori sprigionati dalle torce, l'ebbrezza e le

---

[67] Gabriele La Porta, *Orghia l'orgia sacra a Dioniso*, in Abstracta, a. I, n. VI, Roma 1986, p. 12.

La cresta del Monte Parnaso.

varie sostanze assunte producevano conseguenze estatiche che i devoti incanalavano e sfogavano nella *corsa sacra*.

Secondo alcuni questa poteva tradursi in una caccia rituale e relativo *sparagmòs*, dilaniamento e divoramento delle carni crude delle prede. Ma la stessa corsa, prolungata, poteva rappresentare una forma di possessione in cui la bestia rincorsa si allineava all'adepto, vi si unificava, lo possedeva in uno stato allucinatorio in cui era possibile riscoprire la propria natura più istintuale. A comparire, all'apice visionario, poteva essere lo stesso Toro sacro a Dioniso. Preda e predatore, uomo e animale divenivano una cosa sola e caccia e sacrificio erano ricondotti alla medesima funzionalità, fatto raro nella cultura antica.

In un periodo più tardo le danze e le corse avrebbero preso un carattere maggiormente "commemorativo", imitando le azioni delle mitiche menadi che accompagnavano il dio e la cui natura umana veniva occupata, anche per intere giornate, da quella selvaggia della divinità. Ma non sappiamo appunto se al tempo in cui scriveva Euripide si cercasse ancora di percorrere un'alterazione estatica o se il rito si limitasse, almeno nella maggioranza dei casi, a una forma di rappresentazione, una sorta di *reenactment*. Senz'altro i rituali ellenici giunsero ai nostri Etruschi dopo molte revisioni e adattamenti.

Tuttavia, Pausania nel II sec. d. C. sostiene che a Delfi le donne salivano ancora sulla vetta del Parnaso[68] e Plutarco riferisce, poche decadi prima, che le ritualità provocavano alterazioni della personalità. Racconta inoltre che un gruppo di donne si persero nella *corsa* durante una tempesta di neve e fu organizzato un gruppo di soccorso per salvarle: per poco non morirono tutte congelate, i loro panni divennero «duri come il legno».[69]

L'*oreibasia* può esser considerata, dal nostro punto di vista, un caso estremo di una *possibilità* che l'ebbrezza ispira, conservandola segretamente, e che può manifestarsi in attività, tonalità e intensità diverse. Può aver a che fare con uno "scarico di tensione" che in tempi odierni implicherebbe comunque la danza, il canto, il sesso, ma raramente legati ad una spiritualità iniziatica; mentre in antichità quest'ultimo aspetto si animava nella *choreia*, nel rito orgiastico, nello *sparagmòs*. Altresì la *processione* con la seguente *corsa* - che pare la più essenziale di queste modalità di "smaltimento" (certo depressa nell'intensità e nella portata sacrale) - potrebbero trovare il loro analogo in un *percorso* realistico o metaforico di cui cercheremo di occuparci.

La sinistra attrattiva dei culti-iniziatici trova forse qualche riverbero nell'ebbrezza odierna e, tornando a noi, pure la viticultura etruscofila ha alluso, qua e là, alla loro intrinseca *centratura*. Oltre al potere espresso dai nomi di certe divinità sembrano intuibili, nella complessità dei *blends* e nelle articolazioni del

---

68 Pausania, *Periegesi della Grecia,* X, 4, 3.

69 Da Eric Robertson Dodds, *I Greci e l'irrazionale*, Nuova Italia, Torino 1978 *cfr.* Psiconautica, Misticismo enteogenico, *La oreibasia,* in www.psiconautica.in

lavoro sui vitigni autoctoni, le tensioni di un *metodo* che vuole riconnettere a fattori di profondità; il vino stesso è una sintesi e un'interpretazione del territorio che "ospiterebbe" l'*oreibasia*, e in un certo senso già lo *percorre*, con scelte prettamente enologiche e non solo - nell'anima del vignaiuolo come in quella del bevitore. Ma è forse nell'indagine sul "mito" che si crea e rifonda attorno a un vino di successo che potrà materializzarsi qualche fantasia in più.

Senz'altro dopo il nostro *excursus* nell'antichità si avrà un'idea più consapevole di *cosa potremmo aspettarci* da quel *numen* titolare e protettore della bottiglia, che talvolta ci osserva dall'etichetta nascosto dietro qualche rassicurante sembianza (naturalistica, mitologica, medievale o rinascimentale) e che sembra chiedere come d'essere "seguito". Il nostro Lucumone, pur con la sua autorevolezza e il suo complesso fascino, può "perdere il senno" come il re scita del racconto erodoteo; e il suo sereno *incedere* può divenire *corsa*. Così, questo esperto e distaccato degustatore, celerebbe nell'animo la possibilità di un'espressione selvaggia e misteriosa, tesa al ricongiungimento con il brulicare della natura.

III

Appendice enosofica

OSSERVAZIONI SUL *BERE* CONTEMPORANEO

In Italia, in Etruria e in tutte le terre famose per la propria produzione vinifera, il vino non è solo una componente della scena del convivio o una parte del *paesaggio* culturale, ma è altresì uno dei protagonisti della realtà a cui partecipiamo. In certe occasioni speciali il vino appare come una sorta di ospite, *persona* tra le persone, invitato o auto-invitato per svelare una perenne "buona novella": qualcosa di speciale che ci unisce e che è al fondamento del nostro perpetuo ritrovarci, nella collettività ma anche nella solitudine individuale.

Il vino "siede" quindi a tavola con noi, *dice la sua*, *si fa bere* e al contempo *brama il possederci*, rafforzando un circuito d'affetti che è al fondamento della comunità del sangue, del «consorzio di allevamento umano», della *famiglia* in senso universale. Di quando in quando, la bottiglia stessa sembra avere un'anima: precipitazione nella contingenza eno-gastronomica di un'identità briosa, boriosa ma regalmente bendisposta. Più indefinita di un libro, di un'opera d'arte o di un oggetto da collezione eppur tangibile nella presenza quanto chiara all'intuito nella sua valenza, la vitrea *confezione* somiglia, per certi versi, ad un fenomeno "editoriale" polarizzato attorno ad una figura di autore ideale o di patrono simbolico.

Chi è "seduto", idealmente, al nostro tavolo? Spesso, come abbiamo già constatato, il rimando è a figure del passato storico o ad antiche divinità; i nomi stessi dei luoghi dove si svolge la produzione ad una fluttuante *esperienza personale* e il toponimo stesso si fa «entità guardiana»; si tratta di maschere di un substrato collettivo a cui noi e la nostra anima conferiamo energia.

Il *marketing del vino*, consciamente o inconsciamente, si è da tempo appropriato di questo aspetto e si avverte come, chi produce e commercia, abbia integrato certe possibilità di suggestione. Produrre e confezionare un vino significa presentare "personaggi" con cui è possibile identificarsi o che si rendono *ascoltabili,* insieme al proprio *racconto,* attraverso le immagini, i testi, il gusto, filtrati dall'ebbrezza e non solo. Tutto questo avviene nonostante il *plot* resti spesso vago - il personaggio, appunto, trasfiguri generalmente nella leggenda o in un gossip dalle tinte metafisiche - e raramente l'idea si evolva in narrazione - arenandosi invece sull'umida battigia del gusto.

Prendiamo ad esempio tre vini tratti dalla sfera chiantigiana, così rappresentativa di certe tendenze e coincidenze: il *Gotto di Gottifredo* di Monterinaldi, il *Bennone* di Coltibuono e la *Contessa di Radda* di Geografico. Tutti e tre sono titolati a personaggi storici del medioevo locale, quasi come se necessitassero di un loro "spirito" legato con forza al toponimo attinente alla fattoria. Gottifredo fu un signore di "nazione longobarda" vissuto a Monterinaldi mentre il Bennone altri non è che il Beato Benedetto Ricasoli che consumò le sue penitenze a Coltibuono; non lontano troviamo Radda, con la sua Contessa dai nebulosi contorni storici.

Considerazioni e intuizioni in breve: sembra che il Chianti, qualora decida di alzarsi dal suo trono monolitico-meditabondo per giungere alla tavola del semplice-pasto, abbia necessità d'identificarsi con chi idealmente, di quella tavola tutt'altro che *rotonda*, fu il padrone. L'arcigno longobardo, il monaco egocentrico, la contessa senza pace, tutti presi a tenersi il loro posto a capotavola e guai a toccare le loro vigne, il loro vino, perché «o si fa il santo vino, o si muore»... Tra questi tre Signori, messi insieme un po' a forza, tra sfondi di frutti di bosco e note di pepe (tutti presi nei *dis-arrangiamenti* di una moderata ebbrezza) il

Gotto, con i suo piglio maggiormente *secco*, sembra occhieggiare alla psico-regione degli aperitivi...trasferendo il suo tavolo a bordo pescinale. Si crea così un ponte con quei signori che oggi come ieri, passando da colle in colle, d'aia in taverna, cercano di strappare sino all'ultimo brivido del giorno. Nel frattempo il Bennone, così alto e "chiantico" pur nel suo essere un «vino da pasto» sembra meditare il ritiro in abbazia ed anche la Contessa, stizzita in qualche sua crisi tannica...finisce col rischiar di sfuggirci, adombrandosi in qualche torre raddese. Ma insomma, con questi vini siamo finiti a parlarci, come fossero persone e con il rischio di darci del Tu!

Qualcosa di analogo è avvenuto a proposito dei nostri vini etruschi, spesso definiti da un nome proprio, intriso oltretutto dell'*accelerante* arcaica-mitologica: un misto di divino e d'umano che diverge dalla cronologia realistica spingendoci ad un livello archetipale. Eppure anche questi Fufluns, Thesan, Philika, Heba, sembrano presiedere al nostro banchetto o noi al loro; sono ospiti ed *ospiti*, ed il bevitore, come l'iniziato, sente di poter assumere su se stesso le responsabilità di un colloquio immaginario. Gli *io* che condividono il tavolo, presenti in modi diversi, risplendono come i gioielli di Turan; l'ebbrezza illumina attraverso il volo di Cautha...e così via.

Dunque noi condividiamo la *verve* di questi signori e divinità, *accarezzando la possibilità* della partecipazione alla leggenda che ci è narrata e che pare prossima quanto distante, sfuggente quanto mitica o *sempre valida* (e riportabile quindi alla storia personale). Esperienza del bere ed esperienza storico/leggendaria paiono addensate tra loro e l'ebbrezza crea un sensibile incardinamento. Per parlarne, viene dunque spontaneo rifarsi ad un più generale concetto di *vino-mito*, che comprenda dinamicamente le qualità del vino e le qualità dell'intreccio del *racconto storico* a cui esso riferisce (nomi, immagini, eventi...tutto ciò che va al di là dell'esame del gusto). Insomma, la *sfera* di quel vino in un "*corpus* unico". Ripartendo da questa unitarietà vogliamo aprire ad un'indagine libera e consapevole, orientata all'antichità nella comprensione del contemporaneo; per metter a fuoco come e perché il riferimento al passato possa agire sul presente sino a mimetizzarsi completamente in esso: come quel *Tinia* possa esser ancora "annidato", per così dire, in ogni vino ben riuscito.

Oggi capita di sentir dire che promuovere il vino significa anzitutto raccontare delle belle *storie*. Ma venendo al dettaglio noteremo come difficilmente, un *vino-mito* che non permetta l'identificazione con un personaggio preciso (che sia il losco signore longobardo o il barone ferrigno, la dama rinascimentale o la schiava etrusca, sino al geniale enologo, l'ardito vignaiolo, il personaggio *vip* legato ad un tal vino...e così via) possa coinvolgere il bevitore in un'esperienza enosofica completa.

Il bevitore stesso sembra talvolta partecipare alla leggenda attraverso una specie di *sospensione*,[70] somigliando forse ad uno spettatore cine-televisivo rivolto verso uno schermo interiore: avvinto da una qualche ebbrezza e a tratti ipnotizzato, velatamente curioso o lucidamente invidioso; un po' come il fruitore di una *serie tv*, o pure come un bimbo che può affezionarsi ad un *cartone animato* e quindi divenire dipendente dal suo *merchandising*.

Chi ha lavorato nel settore del *marketing* televisivo sa che una storia in cui tutti i personaggi hanno la stessa importanza, in cui non sia in evidenza un protagonista, subirà un futuro incerto nell'adesione emotiva e quindi nella mera commerciabilità. Ora,

---

[70] Mi riferisco alla *sospensione dell'incredulità* che, secondo i teorici della psicologia del cinema, permette allo spettatore d'identificarsi con i personaggi e "credere" al racconto. Nel nostro caso non abbiamo il *film* ma esiste il *fantasma* di un'intenzione autoriale: l'immaginazione è suggerita artificialmente dal progetto del *vino-mito* e prende energia grazie ad un'ebbrezza indotta.

continuando l'analogia, se è vero che la *serie* di un *cartone animato* non è altro che una complessa episodica pubblicità il cui scopo è vendere i relativi prodotti (giocattoli, vestiario, figurine, gadget, servizi...), nel caso del *vino-mito* possiamo forse concepire la bottiglia come il gadget e la storia di quel personaggio come il film-pubblicità, ridotto però - generalmente - allo spettro dell'immaginazione. In effetti si tratta comunque di una *fiction*, alimentata - specie nel caso della Toscana/Etruria - dalla storiografia stessa oltre che dalle evidenze (paesaggistiche, archeologiche, ecc.) dei luoghi dove si trovano le fattorie. Ecco come può strutturarsi e prendere senso il *vino-mito*.

L'analogia, forse un po' azzardata, inquietante ma a mio avviso plausibile, ci ribadisce che il vino è considerabile una *merce* e che in fondo, chi lo produce, può limitarsi a fare "il compito" bene, allacciandosi attraverso dati sensibili e intelligibili (quelli relativi al gusto e alla visibilità) alle tradizioni che ci legano al territorio (ovvero lo scenario della *fiction* storiografica/leggendaria); e l'impressione è che con il vino, scremando le sovrastrutture, molto spesso si realizzi questa tendenza. Tenersi fuori da tale complesso oggi non è semplice, ma senz'altro è possibile affrontarlo con consapevolezza e dignità senza cadere nel banale e meramente "commerciale".

Conveniamo che il legame tra vino e racconto, tra trame sensibili e trame storiche, mediato dalla bottiglia e dal suo *marketing*, sia evidente nella sua semplice possibilità: la suggestione - intesa anche come semplice suggerimento/sussurro del racconto - è alla base del fascino del *vino-mito* e la sua forza sta nel rilanciare una situazione *possibile* o che *è stata possibile*; lungi dal dramma dell'attualità e in questo fluttuante, immaginaria, attiva in un continuo interscambio con l'universo personale e la sua psichica episodicità. Quella essenziale al vino, è una *serie* a cui contribuiamo noi stessi, con i materiali della nostra fantasia -

anche se non sempre saremmo capaci di ri-raccontarla in modo consapevole. D'altronde la *sfuggevolezza* è qui un carattere paradossalmente costitutivo, ereditato direttamente dalla *terra*, dalla natura e dal suo figlio prediletto: *Dionysos*.

La temporalità del vino-mito

Questi aspetti di non-sostanzialità sono al contempo favoriti dall'impossibilità di *registrare* e quindi conservare le esperienze gustative: tornando ancora all'analogia televisiva-editoriale, notiamo come tanti altri prodotti (con immagini, suoni, testi) siano registrabili e riproducibili mentre è praticamente impossibile far qualcosa del genere con i *sensi* implicati nel consumo del vino. Difatti l'esperienza del gusto non può essere fermata e riprodotta in un "fossile" come si fa ad esempio con le riprese video, audio o fotografiche di una medesima esperienza. Il vino coincide con la sua stessa riproduzione e non ammette mediazioni di formato: per rifare l'esperienza dobbiamo ri-berlo; la bottiglia è sì contenitore, ma non semplice formato di riproduzione poiché i riproduttori, in un senso più esteso, siamo noi stessi con la complessità biologica dei nostri sensi primitivi e con una profondità psichica che non ammette golem tecnologici. E di questa riproduzione siamo pure gli spettatori. La bottiglia, in senso più concettuale, è contenitore di un fantasma naturalistico, di qualcosa/qualcuno d'ulteriore che *sfugge* alla sostanza.

D'altronde non potrò mai comprendere, che so, la *mitica annata* di un vino prodotto decadi or sono, nemmeno se avessi conservato la bottiglia originale...perché nel frattempo il vino invecchiando sarebbe cambiato. Il vino, finché ce n'è in cantina e finché dura, può esser *riprovato*, ma non *riprodotto*. E per esser riprovato in abbondanza e in *serialità* deve assoggettarsi ai cicli della vendemmia, con risultati mai identici o definitivi. In pratica, per *essere*, il vino necessita d'aprirsi - seppur a tratti e in modo a volte assai semplificato - all'istante della realtà animica del bevitore; e questa trova nel *vino-mito* l'*habitat* per rimanifestarsi o riproiettarsi, pure in senso narrativo.

Così anche se riproducessimo, per assurdo, un ipotetico *nastro* del gusto (continuo l'analogia tecnologica), questa resterebbe, assai più della semplice riproduzione d'immagini e suoni, un'esperienza soggettiva; ciò sembra dipendere dall'importanza che odori e sapori prendono ancor oggi nel riconnetterci alla nostra base arcaica e mai definitivamente codificata; ma forse anche perché in altri "media" non esiste, o è assai limitato, un effetto psicotropo-empatogeno. Insomma il gusto, avviandosi e introducendo, "riaprirà" una sfera percettiva e speculativa a cui diversi individui daranno diverse risposte e interpretazioni, un Eden dove ciascuno darà ai *frutti* (e ai loro *sentori*) il nome che preferisce.

Ma l'analogia anche in questo caso ci suggerisce alcuni spunti interessanti e forse condivisibili. Proprio come se esistesse un *nastro* che riproduce un racconto, la trama che coinvolge quel personaggio - il *personaggio* del vino – sembra a volte ricominciare a svolgersi, allo stato di *rêverie*, catalizzata dalle suggestioni del gusto a cui è connesso e dall'accrescersi dell'ebbrezza. Mentre queste fantasie si rendono disponibili nella profondità individuale, a un livello più superficiale le speculazioni sul vino, i "gossip storici" e in particolare il sondaggio della degustazione, presentano comunque una loro speciale temporalità e potrebbero corrispondere a sentieri che, seppur distinti, attraversano idealmente un territorio analogo a quello del mito (già posto in vaga comunione con il territorio di *produzione*), intrecciandosi e seguendo l'eroe nel suo percorso – l'entità guardiana che lascia la sua torre e i suoi cipressi, il compagno di tavola ideale che, all'amplesso del convivio, si alza e parte per la sua *promenade*. Il vino, oltre ad avere un fantasma personalistico, sembra suggerire l'esistenza di un "fantasma di temporalità" che spinge in noi per svolgersi e *serializzarsi*.

Del resto, una temporalità naturale-archetipale, propria del racconto mitico e dei "*nastri universali*" dei cicli cosmici, può esser considerata all'essenza della genesi del vino stesso. Dietro le suddette entità vi è forse, ancora una volta, lo spirito di *Fufluns/Dionysos* che li *possiede*, che *compare e scompare* con rapidità e immaterialità; che si può incontrare solo "dopo aver simulato di perderlo"[71] e che può tornare improvvisamente dall'amica Persefone/*Vesuna*, dalla madre *Semle*, dall'amante *Areatha*, comunque in un *infero* terrestre che si pone al di là della cronologia realistica. Attraverso Dioniso tutti i *vini-miti* annettono e celano un accesso infero: un "baratro" individuale in cui le pulsioni fondamentali possono svelarsi, e le figure e i fantasmi personali cercano di ridisporsi attraverso i disegni dei nostri piaceri e delle nostre nevrosi: è forse questa loro stessa ridisposizione, alimentata dal tenore alcolico, a delineare, "srotolare" il territorio ideale-immaginario del *vino-mito* e la sua temporalità.

Ciò che ho definito, non senza qualche giocosità, *nastro* è ciò che contiene e "formatta" l'esperienza ideale dell'eroe del *vino-mito* nella sua riproduzione favorita dall'ebbrezza. È il *corso* stesso del tempo del mito legato a quel vino e la sua sostanza è inevitabilmente fantastica-immaginaria. La possibilità di questo svolgimento, un tempo parte costitutiva - attraverso il rito - delle conseguenze della fruizione del vino, può essere indagato più a fondo nelle sue tracce percepibili nell'ambito della *bevuta*. Parlare della temporalità di tale esperienza significa anzitutto

---

[71] «Alle feste *Agrionie* di Cheronea le donne lo cercavano invano, e ritornavano con la notizia che il dio era presso le Muse che l'avevano nascosto.» M. Eliade cit., p. 401, *cfr.* W. Otto, *Dionysos*, p. 79.

evidenziare la *diversità* del suo statuto d'essere, distinguendola dallo svolgersi di un *bere* convenzionale o banalmente vizioso. Uno statuto che ha a che fare con una «prospettiva infera» e quindi con una tendenza ad aprire un varco per le profondità della nostra psiche.

Ora, si potrà avere l'impressione d'occuparsi di vino in modo sin troppo generale e forse con vaghezza, perdendo di vista la specificità dell'oggetto di questo libro. Ma in una ricerca sulla metafora di un *infero* o un *aldilà* del vino, come potremmo trascendere dai vini titolati a *quegli Etruschi* che più d'altri popoli risultano legati alla dimensione *sepolcrale* e a una coesione visionaria tra mitologia e realtà? Ad essi torneremo inevitabilmente entro breve, ed è grazie a loro che ci troviamo qui.

Torniamo dunque, per un attimo, a guardare al passato. Considerando che il vino esiste ed è consumato da molti secoli prima che il *marketing* facesse proprio il meccanismo mitologico, viene da chiedersi che ruolo e "peso" avesse il *vino-mito* in epoca remota. Come si manifestava ad esempio nel medioevo, quando la *storiografia* non affascinava allo stesso modo di oggi? Inoltre, come possiamo distinguerlo nell'esistenza di quegli antichi che vivevano ancora *immersi nel mito*? Come intendere insomma ciò che abbiamo definito *vino-mito,* nel tempo stesso del mito?

In antichità, anche quando il bere era semplice e conviviale, non si poteva dimenticare che i "maestri", quelli in contatto con gli dei, svolgevano la loro *unione* attraverso certe ritualità. Il vino era *di Dioniso* anche se il dio poteva essere invitato a presenziare in modo più o meno evidente. Oggi invece la divinità, l'eroe, la ninfa, il santo, rappresentano solo una parte della "sacralità" del vino e non esiste una vera e propria reverenza (salvo ovviamente, ma in tutt'altro contesto, nel caso del rituale cristiano). Il rapporto con i personaggi leggendari è ridotto spesso a una sorta di gioco intellettuale, di cui si può a tratti sorridere, mentre un tempo, l'incontro, anche se talvolta ditirambico, avveniva probabilmente in piena sacralità e col relativo *thambos*. Là fuori vi era il Parnaso o il bosco di Nemi e si sapeva che, attraverso un processo mentale e rituale, si poteva intraprendere un percorso reale quanto estatico-spirituale. Il territorio era portato a coincidere con il *vino-mito* stesso e assumere una più dilatata temporalità. Oggi questa possibilità non è percepita e pare comunque poco riconoscibile. Che se ne dica, il territorio è vissuto in modo letterale (in senso *storico* o *produttivo*); non necessariamente in senso materialistico ma, certo, tutt'altro che metaforico. Esso non corrisponde a un organismo vivo e brulicante che custodisce divinità erranti, anche se il *marketing* del vino, con i suoi nomi e le sue etichette, vuole spesso suggerirci questa impressione.

La degustazione come rito preparatorio

A volte sembra che la temporalità del *vino-mito* sia come inceppata e che per andare oltre, provare qualcosa di più autentico, si debba prima procedere ad un improbabile *riavvolgimento*, riconsiderando i nostri gesti iniziali e la nostra disposizione d'animo. Proviamo quindi a ricominciare dall'*incipit*. A pensarci, in antichità tutto cominciava in modo convenzionale, serioso ma semplice e ancora oggi noi attuiamo gesti introduttivi che provocano rispetto e quasi riverenza, anche se non procediamo ritualisticamente al di là di essi.

Forse, essenzialmente, la degustazione relativa ai primi sorsi non corrisponde altro che a una preghiera: la lettura dei *caratteri*, del *liber* di quel vino; il riepilogo delle sue *regioni interne* e del programma del loro attraversamento; avvenendo, idealmente, prima di ciò che un tempo fu la *craterizzazione*, si pone all'inizio di un rito possibile anche se l'obiettivo resta quello di «riconoscere» il vino mantenendosi analitici e letterali.

In qualche modo, degustando noi visitiamo già quelle *regioni interne*, volando alto o pedinando i protagonisti del racconto mitico; avanzando come "in incognito" nei cespugli, scrutando tra i rami, raccogliendo *sentori* e *sapori*. I più bravi riescono a determinarvi qualcosa di oggettivo, descrivono il sensibile come il leggendario ma sappiamo che quell'*aldilà* è più grande; che una figura si muove *oltre la macchia*; e che una parte di noi vorrebbe e forse potrebbe essere *con lei*. Questo *riepilogante pedinamento* lo compiamo, per così dire, in uno stato fantasmatico e non siamo protagonisti della circostanza, come avverrebbe in un rito o in una coazione realistica.

Senz'altro, volendo interpretare certi momenti e situazioni dal punto di vista di una "mancata *oreibasia*", dovremmo concentrarci su aspetti più sottili o mimetizzati, forse ammantati di banalità e che riguardano probabilmente la seconda parte del convivio e forse il momento conclusivo, in cui ci si alza dal tavolo o si ipotizza di farlo; e, volendo, ciò che può succeder dopo.

L'impressione di un percorso rituale svolto all'interno di un territorio sacralizzato, ora suggerito dalla storia e dalle tradizioni - filtrate da sensi, parole ed ebbrezza - può manifestarsi, durante il pasto e/o la bevuta, come spunto immaginativo o fantasticheria intima del bevitore. Parlando con molti apprezzatori di vino, ho potuto constatare che è spesso riconoscibile un momento in cui una "chiamata" si fa percepibile alla coscienza;[72] in essa risalta un complesso legame tra il fattore alcolico e gli orizzonti del dialogo (intrapreso con se stessi o con i compagni di bevuta, quindi in senso più empatico). Ma si può percepire un'apertura in senso più generale, intesa come parziale caduta delle "barrie-

[72] Si veda anche *Il Quarto Chianti, cit. Aggiungerei che questa fase può coincidere dal punto di vista fisiologico con la diffusione dell'etanolo nell'organismo e dal suo intervento nelle attività cerebrali; ma a interessarci è il significato che questo "momento" acquisisce declinandosi al vino-mito.*

re" del *cosciente*; qualcuno ad esempio ricorda improvvisamente i sogni della notte precedente. Comunque, aperto un varco, capita d'intuire l'idea di un "messaggio" proveniente da "fuori", se vogliamo dal "territorio".

Questa "chiamata" può esser colta tanto con estro intellettuale ed ironia quanto materialisticamente, con conseguenze assolutamente realistiche, in seguito persino coattive o "triviali" - soluzioni che al momento non ci interessano.

Ma perché l'intera esperienza si riveli avvolgente, coerente e consapevole, l'esistenza di un momento di passaggio all'azione dovrebbe essere già riconoscibile nel "raccoglimento" della degustazione. In pratica la degustazione dovrebbe farsi "profetica" di una successiva *oreibasia* o di quel che oggi può, almeno in parte, sostituirla; ciò passerebbe dalla consapevolezza dell'avvento dell'ebbrezza e quindi dal riconoscimento della gradazione alcolica come possibile fattore sinestetico.

La ricerca di un centro

Quel che segue ora è, forse più di prima, una speculazione personale, non priva di una praticità che può portare lontano e che chi legge potrà apprezzare o no secondo le sue predilezioni; vi è inoltre l'intenzione di testare altre analogie, per la curiosità di vedere dove ci conducono.

Tornando a noi e alle circostanze attuali, diciamo che persino una semplice passeggiata "per smaltire" il pranzo appena consumato - ad esempio nei sentieri collinari nei pressi di un ristorante di campagna - può prendere un senso vagamente iniziatico. In tali "girate" potrebbe rivivere il germe sopito, il *demone* (parzialmente *disinnescato*) dei percorsi delle antiche processioni, *corse*, riti.

Queste *promenades* puntano solitamente a obiettivi conosciuti: piccoli monumenti, luoghi panoramici, fontane, cappelle, alberi, boschetti, giardini, panchine; o anche ad un particolare bar, una piazza, un caffè (i convivi si svolgono spesso in luoghi da cui si possono intraprendere delle piacevoli camminate che risultano così ancor più spontanee). Già in queste consuetudini possiamo riconoscere l'idea di un percorso catartico, un'attività reale che accarezza il corso di un'esigenza più profonda. Corsa, sesso, caccia e sacrificio sono ovviamente esclusi, e quel che resta nell'immaginario fa parte di uno scenario interiore dove viviamo, tutt'al più, l'incontro con le nostre immagini a loro volta sovrapposte a funzioni collettive, se non al *logos* intrapreso con chi ci accompagna.

La possibilità di svolgere questi cammini "iniziatici" può appunto manifestarsi già durante il pasto, in un particolare momento d'evoluzione - un passaggio da lieve euforia a piena ebbrezza - in cui «si decide» il senso della nostra partecipazione...o è il *demone* a decidere per noi; forse non a caso, come abbiamo visto, i vini etruschi maggiormente alcolici recano nomi di divinità.

Si procede quindi raccogliendo, intimamente, una vaga promessa estatica, per materializzarla quando la digestione si allineerà all'evoluzione della *passeggiata* e quest'ultima favorirà la riproduzione di un *corso*. Ecco che la nostra anima lascerà scorrere i suoi corrispondenti delle antiche immagini; *smaltire* il pasto significa quindi, al contempo, smaltire le fantasie (a una ricerca più approfondita scopriremmo forse, in questo caso, delle profonde simpatie tra digestione e immaginazione).

Infine, trovato un *trono* col suo paesaggio, un palco o qualche piano geometrico (spesso con riferimenti alla circolarità), chiacchiereremo ancora gestendo l'ultima ebbrezza o continueremo a "lavorare immagini" in solitudine. Forse ritroveremo da qualche parte in noi l'eroe che si era improvvisamente alzato da tavola, o la ninfa che avremmo voluto trattenere. Avverrà insomma un incontro con figurazioni private celanti aspettative, desideri irrisolti o repressi, rimozioni, piccole e grandi nevrosi. E tra queste figurazioni, qualcuna più in evidenza sarà forse animata dai complessi di *Fufluns* o di una delle sue compagne.

Così un'ideale bottiglia "dionisiaca" può recare, nell'etichetta, lo scenario stesso dei colli in cui si svolgerà la nostra moderata *oreibasia* (reale o immaginativa che sia): forse la rappresentazione del sentiero che percorreremo nella *promenad*e, o della fattoria o borgo da cui esso prende avvio; o del monumento, torre, cipresso, laghetto in cui questa avrà la sua meta; o di qualche *visione* che ci coglierà improvvisamente, nel buio, se avremo l'ardire e la fortuna di compiere la nostra "girata" in orario notturno - e magari in un ambiente boschivo/romantico.

Se in etichetta troviamo simboli geometrici, come avviene nel caso di molti raffinati vini toscani e di alcuni dichiaratamente etruschi - come La Rasenna, Porsenna, L'Etrusca o il *Lucumone* cortonese -, possiamo immaginare questi schemi come anticipazioni delle ritualità pertinenti al luogo sacrale ove conduce il cammino: *rings* adibiti all'incontro con una parte di sé, e/o con quel personaggio la cui figura è da subito evocata, nella varietà delle sue sembianze arcaiche, in etichette di vini come Etrusco, N'Etrusco, Rasna, Vel Aules, Manero.

L'etichetta, in un senso contenutistico-concettuale o in un senso grafico-simbolico, è comunque la *quadratura di un cerchio* - quest'ultimo corrispondente alla bottiglia con la portata del suo caos dionisiaco. L'esistenza di scritte che vanno da destra a sinistra (secondo l'usanza etrusca), forse un «movimento in direzione dell'inconscio», evoca l'*incipit* di tale processo ricordandoci una nota interpretazione junghiana: al centro «si dice vogliano ricostruire il *gibbone*»,[73] che in questo caso avrebbe qualcosa a che fare con *Fufluns*/Zagreo risorgente.

D'altro canto, osservare in altri tipi di etichette immagini di luoghi metafisici, talvolta idealizzati o avveniristici, spinge a pensare che gli autori avessero intuitivamente chiara la necessità di creare dei riferimenti, mete, assi o centri. Ad esempio, gli spaccati campestri dove gli animali di Urlari puntano alle stelle, potrebbero essere al contempo teatri di riti silvestri. E non sembra forse, in altro senso, l'immagine della cantina di Hebo la riproduzione di una grande ara o tempio? a pensarci, non è l'unica cantina capace di trasmettere impressioni del genere.

Queste suggeriscono, a chi vuol cogliere, che in fondo al percorso può avvenire un "sacrificio", o un ideale "passaggio". Il territorio intorno al convivio è stato trasformato in un'anticamera dell'Ade, e il ristorante originario potrebbe essere inteso come una specie di "avamposto"...come avviene nei romanzi nel caso di certe taverne o spedali.

Ma la ricerca di analogie potrebbe portarci ancora più lontano. Tuttavia, a proposito del rapporto tra il mito e la realtà del bevitore, siamo andati anche troppo avanti: le stesse etichette sembrano suggerirci che esistono valenze peculiarmente immaginarie e immateriali; mettendo da parte i percorsi e le *promenades* fisicamente intese, riteniamo che la "chiamata" dell'*eroe* possa essere colta in modo prettamente psicologico e che il racconto/*nastro* del *vino-mito* possa riprodursi nell'animo del singolo, o nell'ambito di un *logos* coscientemente condiviso, attraverso materiali colti dal reale (territorio, persone) e implicati nel raggiungimento di un *centro* o di un *punto d'equilibrio*.

Nei casi più lucidi si potrebbe parlare forse di una "medita-

---

[73] C. G. Jung, Opere XII, *Psicologia e Alchimia*, Bollati Boringhieri, Torino 1992, pp. 126 e seg.

zione da ebbrezza" o di una speciale psicanalisi. E il discorso si farebbe oltremodo complesso.

Le qualità dell'ebbrezza

Che si tratti d'immaginazione o di qualcosa di più concreto, è plausibile che i singoli vini (e in particolare quelli etruschi per la loro complessità estetica) abbiano modi diversi e speciali di corteggiare la possibilità della riproiezione di temporalità iniziatiche-rituali. Ma tanto le qualità organolettiche quanto soprattutto l'alcolicità - quindi le caratteristiche dell'ebbrezza - possono avere impatti molto diversi, di bottiglia in bottiglia e, ovviamente, di persona in persona.

Ad agire pienamente sull'integrazione del *vino-mito* sembra essere un "quarto esame" o fase gustativa (ma forse nemmeno definibile più tale). Per raggiungerla coscientemente dobbiamo superare l'idea di *riconoscere le differenze* del gusto e realizzare la possibilità di *riconoscere le differenze delle promesse estatiche*. Probabilmente, con un po' d'ispirazione, quest'operazione potrebbe svolgersi già da subito, appena ci approcciamo a calice e bottiglia, aggiungendo qualche paragrafo al *liber* della degustazione, al "raccoglimento" del *riconoscere*; favorendo l'incontro tra immagini e tenore alcolico attraverso l'intuizione della portata del *vino-mito*, quindi dell'analogia che lega le diverse accezioni e componenti del suo intreccio.

Abbiamo già supposto che la degustazione sia considerabile di per sé una specie di avventura, sebbene sostanzialmente osservativa: un'esplorazione in "incognito" nel territorio del *vino-mito*. Prima si attraversa la sfera visiva e olfattiva, poi quella del gusto riconoscendo sentieri di morbidezza o di durezza, ascese e discese d'intensità o equilibrio e anfratti di persistenza; il profumo inebria e s'intravedono gli dei; si corre tra tannini, acidi e alcoli in cerca della *sophia* di quel vino. Si scova lo stesso *cervus* inseguito da quel Barone, Abate, Contessa e infine, nell'afflato della persistenza, si comincia a immaginare un giudizio sulle qualità. Ed è qui che forse, nel nostro tornare a una visione realistica, sospesa sulla superficie del sensibile e distante da un livello emotivo-arcaico, ci sfugge qualcosa d'importante. Quel giudizio nasconderebbe una pulsione più profonda in cui il fantasma dell'ebbrezza è già pronto all'azione, come il cervo è pronto ad essere immolato. Queste impressioni ci invitano a indagare più a fondo la presenza del *divino* già nei primi istanti di conoscenza di un vino; o forse, con qualche prudenza, il "quarto esame" gustativo potrebbe svolgersi più avanti nella bevuta: al momento di quella che abbiamo evocato come «*chiamata*».

Estendendo la "durata" della *degustazione* ma inchiodandola alla sua naturale vocazione concettuale, qualcuno si sarà chiesto - domanda condivisibile quanto paradossale - perché non cercare di valutare e scombinare le componenti del fenomeno dell'ebbrezza così come si fa per i sentori, l'intensità, la struttura del gusto. Indubbiamente i suoi fattori, che partono da una base alcolica/sensoriale per elevarsi a livello piscologico-esistenziale, prendono valenze difficilmente calcolabili. Ciascuno è toccato in modo diverso dalla presunta *chimericità*[74] di un vino in base al proprio fisico, alle esperienze, all'abitudine a diverse gradazioni, allo stato d'animo da cui *parte*. Inoltre, attraverso l'alterazione alcolica, ciascuno incontrerà immagini diverse e personali che possono perturbare la percezione del fenomeno.

Descrivere in modo oggettivo le qualità dell'ebbrezza sembra davvero un'attività da *maestri*, ma soprattutto da autentici cono-

---

[74] Il concetto già inoltrato ne *Il Quarto Chianti* (*cit.*) definisce l'ipotetica *qualità* con cui i singoli vini possono influire, in modi diversi, sulle responsabilità e le aspirazioni esistenziali dei bevitori.

scitori di sé: del proprio materiale animico e di quel che esso cela; delle proprie potenzialità biologiche attraverso le quali l'alcolicità, corteggiata dal gusto e dal *vino-mito* in generale, viene declinata ad una dignità umana e filosofica. Un po' come un *sommelier* ha un grande naso, un "maestro d'ebbrezza" dovrebbe avere un'abissale e vivida *tensione*. E la prima cosa può escludere l'altra, o viceversa.

Proprio quando un *sommelier* comincia a trovare difficoltà - per comprensibile corruzione degli apparati sensoriali – e dunque quando l'alcolicità comincia a occupare il campo, potremmo riconoscere l'*apparizione* dello scenario del percorso iniziatico: fantasma di quella *possibilità* che, concatenandosi ad altre, si è fatta territorio e trama in cui ambientare la nostra *moderata oreibasia*. Dunque, ecco manifestarsi l'opzione estatico-iniziatica: ora siamo una cosa sola con il demone, ebbro per definizione. Una qualche preda, ninfa, immagine o concetto personale si offre sull'altare del sacrificio; ma il modo in cui si conduce questa *figura* ad Ade, il modo in cui ci uniamo ad essa e i risultati esistenziali di questa soluzione, variano da individuo a individuo, ma anche da vino a vino.

La *tensione* dell'incontro, incardinata sulle qualità dell'ebbrezza, potrà rendere briosi o nevrotici, ispirati o malinconici, e quello stato lo condivideremo solo con una parte degli altri bevitori che per ragioni varie e complesse, umane e biologiche, ci somigliano in qualche senso. Senz'altro, con quel vino avvieremo un rapporto "interpersonale"; lo vorremo incontrare ancora, per capire se la sua *chimericità* è la medesima, se ancora ci suscita le stesse passioni: i rapporti tra bevitori e singoli vini possono essere lunghi e persino travagliati.

Perdendosi nel *territorio* di quel vino attraverso gli esami "iniziatici" del gusto, ed aprendosi un varco nell'*aldilà* di quel vino, conosciamo un *secondo giudizio* di cui il primo era solo un'ombra convenzionale. Difatti, avvicinandoci all'autorità del demone-titolare, compromettiamo al contempo noi stessi, rischiando in certi casi l'imbarazzo, meccanismo complesso e curioso, la cui analisi è senz'altro da rimandarsi ma che presenta già improvvise emergenze paniche durante il corso del convivio: i capri espiatori condannati nelle chiacchiere e nei *gossip*; la tentazione di mostrare parti di sé più molli e nascoste; la stessa penitenza che applichiamo a noi stessi attraverso l'imposizione della *promenade*...oggi conclusa con una misteriosa indulgenza personale, un tempo cardine dell'estasi rituale.

Quasi come se la degustazione stessa fosse il risultato convenzionale della rimozione dell'iniziazione stessa, confermandone forse, segretamente, la struttura riproduttiva.

Una prospettiva "mortifera"

Dunque tra degustazione e immaginazione, tra pasto e *promenade*, l'antico *numen* ritorna e si scompone, frammentato dall'urto della rimozione, in una moltitudine di figure e immagini riesumate dalla storia e dall'inconscio personale del bevitore. E queste, grazie all'estetica dei singoli vini, sono riproiettate nelle entità del *vino-mito*: gli eroi, i toponimi, i sentori, dominati da un demone titolare che pretende e garantisce le medesime analogie, assicurando un ritorno a una base collettiva e arcaico-sensoriale.

L'etichetta si spande e si fa territorio, tutto ci parla di noi stessi ma al contempo di presenze remote; da qualche parte, nel mondo oltre l'etichetta, si anima qualcuno o qualcosa...forse una preda che è una cosa sola con il dio. Il territorio dovrà essere consacrato attraverso il rito, l'estasi raggiunta attraverso la cattura o il gesto balistico: la scoccata di Bellerofonte, il tuffo del nuotatore infero, la rovesciata del funambolo, l'assolo di tromba di Belbo...fino allo stesso *prosit*, l'alzare il *calice biconico* con vigore innanzi al paesaggio stesso. Preparazione, corsa, sacrificio, nutrimento. Riproduzione e morte. È scontato che si abbini carne al Sangiovese: non è solo perché entrambi hanno un sapore "deciso", o altro.

In quale dei colli rossastri che mostra l'etichetta di questa bottiglia catturerò il *cervus* precedendo il *barone* stesso? Perché il colle secerne sangue e perché questo sangue è come il vino? Cosa accomuna, in senso enosofico, il sacrificio *cristiano* con quello degli animali nebrisati e perché tutto ciò rivive in noi, nella carica delle *estasi*? Alla base della legittimazione della "civiltà del vino" sembra esserci un qualche tipo di metaforico spargimento di sangue, immagine intuitivamente chiara ma non semplice da interpretare; abbiamo visto come Fufluns sia legato alla morte ed il *vino-mito* possa essere interpretato come una complessa anticamera dell'Ade. Il vino ha forse a che fare - più di quel che è normalmente pensabile - con una *morte* simbolica? L'allegria di Sileno è come una risposta isterica alla vicinanza al baratro, margine dove il «panico subliminale» della natura si fa quasi ironico, e lo sapevano gli Etruschi che dipingevano con eccitazione i momenti dell'Aldilà – ma non dimentichiamo quel *tuffo* che introduce ad un *nulla* ulteriore, quel *coma* naturalistico, etilico e definitivo.

Ed ecco, tra gli altri aspetti, come può prendere senso la ricerca sui vini etruschi, da intendersi ora come *habitat esistenziale* per la riscoperta di un rapporto - oggi sostanzialmente rimosso - tra vino e oltretomba. Come dire che bere è sempre un po' la ricerca di una Fine...spesso travestita da celebrazione di un *fine* esistenziale. Nel profondo, qualcosa ci invita ad un *aldilà*, accarezzando la possibilità del *taglio* che definisca la nostra esistenza; che ci restituisca quindi ad un Eden ancestrale - forse un caso di speciale mitigazione di un *desiderio di morte*.

Infine sul *paesaggio*, un po' onirico e un po' mortifero evocato dal *vino-mito*, resta difficile affermare parole chiare. Questo paesaggio può essere inteso come frutto di un'immaginazione variamente lucida; ma al contempo ritengo possa esistere una base enosofica-collettiva, legata forse agli archetipi sensoriali che quel vino va a stimolare; e quindi ai sentori, alla struttura, agli equilibri tesi sul perno dell'alcolicità. L'immaginazione e la temporalità di tale *paesaggio* sono drammaticamente soggettive ma animate probabilmente da matrici comuni, orientate *miticamente* al raggiungimento di un *centro*. Vige l'idea di un dualismo intrinseco: ogni vino ha una sua realtà territoriale, umana-produttiva e diciamo, organolettica; ma al contempo ha un suo *aldilà*, un *altrove* che può sfruttare elementi di quella realtà come metafore; nonché una *chimericità* che l'ebbrezza dà l'impressione di scatenare.

Una proposta: il territorio a cui si "accede" potrebbe corrispondere forse un *paesaggio in senso esistenziale*; per capirci, si pensi anzitutto al modo in cui la metafora del paesaggio è stata utilizzata da chi ricerca sulle forme del pensiero; ad esempio quando si parla di andare «al di là» di un certo «principio», o «allargare gli orizzonti», o «perdersi in una selva» di elucubrazioni, utilizzando le immagini di *territorio*, *orizzonte*, *selva*, *itinerario* e molte altre (in questo libro lo abbiamo fatto più volte). Ma ecco appunto che, nel caso si stia ragionando di un vino - forte del suo territorio, dei suoi toponimi e del suo *terroir* - l'analogismo si fa più complesso, ridondante e avvolgente.[75]

Una volta assunte dall'anima le "metafore paesaggistiche" possono, per costituzione, rendersi autonome secondo il linguaggio delle immagini; senza tornare necessariamente a uno

[75] Ho cercato di approfondire l'argomento ne *Il Quarto Chianti*, *cit.*

stato tale da reggere l'analogia con il pensiero che le aveva evocate e strumentalizzate (o che poteva comunque farlo), ma esprimendo "concetti sentimentali" (forse si può parlare di *stimmung*) di autentica profondità.

Questo avverrebbe con più facilità nei casi in cui il *processo primario* diviene protagonista dell'esperienza psichica, e quando si riattiva un livello di elaborazione arcaica: forse prettamente dionisiaca. C'è ovviamente il caso delle attività oniriche (penso ai percorsi/viaggi attraverso fantastici *territori* che capita di sognare nelle fasi di grandi cambiamenti personali), ma possiamo aggiungere le esperienze di alienazione, di alterazione e, ovviamente, di pronunciata ebbrezza.

Nel progredire dell'ebbrezza i *possibili* di una ricerca esistenziale sembrano riattivarsi *germogliando come immagini di paesaggio*; soprattutto qualora il territorio di produzione sia chiamato in causa dal *logos* di una degustazione o di una discussione sul vino, e/o attraverso quel vino. Durante questa attivazione, generalmente intermittente ma sensibile, i nessi associativi sono quelli del *mito* e l'interscambio con l'immagine del territorio può farsi frequente e importante. Del resto noi sondiamo uno spazio mitico, realmente inesperibile, percependolo in quanto flusso di "concetti"; l'analogia tra strutture delle immagini e sviluppi del pensiero (ora lucido e smaliziato) può persistere con decisione e l'esperienza dell'ebbrezza può suggerire, muovendo in chiave archetipica, soluzioni a questioni irrisolte se non semplicemente affermare la loro fatalistica evidenza: eccoci giunti a qualche *centro*.

Questa analogia, nel suo incanto ma anche nella sua perpetua incompiutezza, può essere interpretata da qualcuno, forse superficialmente, come un invito a uscire e cogliere una realtà del territorio. E fu certamente il caso degli uomini antichi e dei loro riti notturni, o di altre consuetudini in cui si agisce con eccitazione, leggerezza, inconsapevolezza. Tuttavia, Fufluns sembra richiamare anzitutto alla profondità della psiche del bevitore. E l'invito dovrebbe forse essere assunto nella propria interiorità come intuizione di un profondo senso del sacro, un'aspirazione alla completezza, un *orientamento* che unisca uomini, terra, sole, vino...

Ma tutto ciò, ovviamente, è questione d'interesse generale: da approfondirsi altrove rispetto a un libro sui vini d'Etruria.

Appunti conclusivi

Nel nostro itinerario di ricerca, prima attraverso la viticultura, poi nella miti-storia e infine nell'immaginazione del "bevitore etrusco", abbiamo più volte intuito l'esistenza di una fantasia iniziatica-sepolcrale che sembra attingere, con tanta vaghezza quanto indubbio fascino, da forme rituali, ludiche e sacrificali di valenza archetipica. In molti casi questa "presenza" è appunto rintracciabile nell'estetica delle bottiglie (immagini, nomi, concetti) e al contempo possiamo ricercarla nello specchio del gusto, spesso gentile e vellutato ma quasi sempre dotato di un ferreo fondamento, derivato dai terreni caratteristici d'Etruria. Inoltre, quando abbiamo applicato una chiave di lettura "rituale" per le circostanze del bere, abbiamo visto realizzarsi lo schema raccoglimento-cammino-raggiungimento che potrebbe accomunare consuetudini realistiche e processi psichici legati da complesse analogie.

Sorvolando su gran parte dei sentieri speculativi che sono stati intrapresi nello scorso capitolo (chi vorrà confuterà o approfondirà a suo modo), possiamo dire che l'esistenza di questa fantasia è plausibile e percepibile, oltre che capace di rafforzare il legame di certi vini etruschi con il proprio territorio – quanto, idealmente, con gli strati ctoni a esso corrispondenti.

Molti sviluppi del vino (produttivi, creativi, degustativi, estatici...) prendono strade peculiari e si localizzano in modi diversi riguardo al baricentro di questa fantasia: spesso lo ignorano totalmente ma, nella loro implicita ricerca di sacralità, sembrano muoversi nei pressi della sua influenza. Del resto, indipendentemente dai nomi e dai luoghi, ogni vino degno di questo nome è in fondo pertinenza di Fufluns/Dionysos, e la fascinazione etrusca resta una "tentazione legittima" sebbene il suo senso iniziatico sia generalmente trascurato.

Difatti, anche se la fantasia è in qualche modo assimilata, rara-

mente s'intende o sottintende le sue implicazioni più profonde che si pongono nell'epoca dei misteri e di una serie di convenzioni dai tratti oscuri, inconciliabili con le abitudini contemporanee. A volte si esalta al contrario la "solarità" della tavola e della gioia di vivere: un fatto positivo e incoraggiante se non fosse che la "gioia" espressa nelle raffigurazioni arcaiche fu generalmente una rappresentazione dell'aldilà, o di un gioco funerario o comunque una celebrazione d'ebbrezza ricca d'implicazioni cultuali.

Che vi sia un fattore rimosso è forse probabile; certi vini etruschi, per così dire, "provocano" un demone, aprono un varco al suo fascino senza poi gestirlo a pieno o ignorandolo – ma d'altronde, oggi, il bevitore è generalmente privo di strumenti concettuali utili in proposito –. Il risultato è forse che la fantasia dionisiaca, con le sue funzioni e le sue figurazioni potenziali, avanza senza troppe interferenze e aderisce più che in altri casi al materiale personale. Così il rapporto dell'io con la natura vinifera, con la materia con cui è chiamato ad interagire, si caratterizza di un'inquietudine che la sepolcralità *etrusca può certo favorire.*

*L'aldilà di un vino si lega alla mutevolezza della divinità e al suo polimorfe seguito di menadi, demoni alati, satiri, bestie, oscuri germogli (*puplu*) di un substrato arcaico. Le mescolanze tra gli uvaggi (spesso a gruppi di tre, come le figure degli antichi specchi) e gli abbinamenti (con il sacrificio della cacciagione e la relativa "consacrazione del territorio") sono appigli per la riproiezione del mito. Ed ogni vino e bevuta etruschi, forse più di altri, conservano la possibilità, seppur remota, di evocare un immaginifico* thiasos.

Così, esistono forse vini-miti *più sinistri e inevasi, che focalizzano maggiormente sull'accezione misterica o semplicemente misteriosa dei territori, della storia e dei suoi personaggi. Come detto, alcuni vini paiono conviviali o semplicemente banali; ma altri presentano scelte più complesse, apparentemente minimaliste, dietro le quali s'intuisce un intenso lavoro concettuale, una preparazione, uno schema, in-*

somma il retaggio di un rito o una "magia" – o così può apparire all'anima, questo lavoro, se circondato dai riferimenti etruschi di per sé enigmatici.[76] *Quei puntini tra le lettere o altri gesti minimi potrebbero sembrare dettagli, ma celano forse tensioni ulteriori; quelle scritte etrusche da leggersi verso sinistra, potrebbero prendere la direzione di un'insondata profondità.*

Insomma: alcune caratteristiche del vino in generale troverebbero negli archeo-vini e specialmente nel vino etruscofilo un caso di manifestazione paradigmatica.

Al contempo la celebrazione, più o meno implicita e psicologica, dell'incontro con il demone etrusco e con il suo varco infero, è probabilmente la forza, così come il limite, del marketing dei vini che si rifanno dichiaratamente agli Etruschi. Forza, per il notevole fascino che abbiamo visto più volte scaturire. Limite, perché a una parte dei bevitori questo "ritorno agli inferi" parrà sempre inopportuno, oltre che vetusto - più che antico. E in tal senso, scelte come quella dei caratteri antichi o dei riferimenti alle pitture vascolari non aiutano a superare le esitazioni, anzi tingono il contesto di un qualche esoterismo.

Nonostante ciò il vino etrusco non è da ritenersi un vino per intellettuali o per bevitori "selezionati": è per tutti coloro capaci d'intuire, anche inconsapevolmente - cioè con la semplice infatuazione - la portata di un richiamo*; per chi riesce a percorrere una qualche ombra e danzarvi insieme come in un affresco sepolcrale; per chi oltre i paludamenti delle etruscherie riesce a intravedere una freschezza primigenia.*

...È plausibile che questa autenticità si esprima in alcuni Etrusco o Rasenna, in Turan, in Philika, in Zilath, così come nelle immagini di molte fattorie; più complessi paiono i casi dell'affascinante Vel Aules, del ferrigno Rasna, dei suggestivi Camars, Pesna, Heba,

---

[76] Tali aspetti tendono forse ad evidenziarsi quando ci allontaniamo dai nomi più generici (*Etrusco ecc.*), preferendo definizioni concettuali o di specificità toponomastica.

Aska...o di altri che riferiscono alla complessità di Voltumna e del capo degli dei (Tinia). Mentre per Cautha, Thesan, per il pregnante Fufluna e per qualche Lucumone, sembra che quell'energia interiore, quell'aurora che si cela oltre il "filtro oscuro" di un passato rimosso o perduto, sia evocata con un certo coraggio.

Dunque concluderò con un'ultima intuizione riferita ad un possibile orizzonte di questa ricerca.

L'antico demone etrusco è ciò che il bevitore può incontrare ma che il vignaiolo, di fatto, potrebbe favorire, o per così dire evocare. Questo richiamo è forse presente nel "mistero enosofico" di altri famosi produttori e non possiamo negare a priori che qualcuno di essi abbia cercato di far propria una valenza misterica nella sua ricetta oggettiva, cedendo almeno per un breve tratto alla "letteralità" del richiamo. Ma come?

Potremmo forse guardare all'etrusco aretino...a quell'uomo amabile, dai gesti spontanei, che benedice i vigneti con i suoi compagni àuguri e che tiene accesa la musica per i suoi vini [I.]. Ma forse, nell'immaginare una specie di consacrazione, abbiano in mente qualcosa di più complesso.

Parlo forse di un opus *improntato sulla produzione del vino? Di qualche formula che va oltre il normale utilizzo di una vigna o di una cantina? O di un geniale metodo di "vinificazione alchemica"? O di una prescrizione per sfuggire alla* mimesis *della grande distribuzione? O di un marketing di natura complessa e superiore, incomprensibile pure a certi "esperti"?*

Pare si stia fantasticando ma d'altronde il vino, e in particolare la creazione del mito di un vino, resta cosa da "maghi" e "aruspici"...è un evento mai del tutto chiaro, e questo è il bello. Cos'è successo realmente in quel casolare, in quella cucina o in quel laboratorio? Chi o che cosa ha ispirato quelle sorta di geni? Un qualche segreto sembra aleggiare nel silenzio di certe cantine, nel mito che vi si vuol costruire intorno, tra le righe della storia che quel vino-mito *è – stavolta attraverso la figura eroicizzata degli stessi produttori - ultimo stadio possibile dell'evoluzione personalistica del vino-mito stesso -.*

Quelle dolci colline dove si manifestò il demone, che sono le stesse famose in tutto il mondo per il loro vino; e quello spargimento di sangue (i conflitti antichi e medievali, l'indagine dei fegati delle bestie sacrificate, e non solo) il cui ricordo sembra evocare una misteriosa comunione ontologica con la spremitura e il consumo del vino stesso; ma anche quegli abbinamenti di selvaggina e d'immagini, di uvaggi e dreifigurengruppen, *di consuetudini e scaramanzie grafiche, di strategie di marketing...corrispondono ad un medesimo intreccio, intimamente ed originariamente coinvolto nell'immaginario d'Etruria.*

RINGRAZIAMENTI

Oltre a ricordare tutti i compagni di bevute e di ricerca, devo ringraziare Enio Pecchioni, mio padre, che è stato indispensabile per le parti storiche.

Ringrazio inoltre tutti i produttori, quelli che mi hanno accolto e quelli che non sono riuscito a raggiungere; i tanti di cui ho potuto bere il vino ed altri per i quali ho dovuto usare un po' d'intuito. Grazie per lo sforzo creativo in questi tempi così complicati, e per le "bevute" che ci hanno scaldato il cuore anche in momenti difficili.

L'autore

Indice

Finito di stampare
nel dicembre del 2017

Made in the USA
Monee, IL
08 November 2026

67968600R00097